Geospatial Analysis of Public Health

Gouri Sankar Bhunia · Pravat Kumar Shit

Geospatial Analysis of Public Health

Springer

Gouri Sankar Bhunia
Department of Science and Technology
Bihar Remote Sensing Application Centre
Patna, Bihar, India

Pravat Kumar Shit
Department of Geography
Raja Narendra Lal Khan Women's College
Midnapore, West Bengal, India

ISBN 978-3-030-01679-1 ISBN 978-3-030-01680-7 (eBook)
https://doi.org/10.1007/978-3-030-01680-7

Library of Congress Control Number: 2018960199

This Springer imprint is published by the registered company Springer Nature Switzerland AG
The registered company address is: Gewerbestrasse 11, 6330 Cham, Switzerland

Dedicated to beloved teachers and parents

Preface

Medical geography is incredibly a dynamic sub-discipline of geography which is conventionally related to the spatial aspects of disease ecology and healthcare management. With the expansion of geospatial technology, medical geography has been transformed to formulate several measurements from far above the earth surface and create dozen of maps of disease and health events within a short period. In this book, we look forward to achieve geographical aspects in public health research and analyze geographical distribution of population exposed to threats and health outcomes and to tackle public health problems. This book will be supportive in providing a blueprint of the dimensions of spatial distribution of diseases and the associated environmental health control measures.

The book has been structured into seven well-organized chapters. The introductory portion of this book will contain data collection, data organizations, data standardizations, and the description of the complications innate in interpreting semantics. In an effort to provide some common background in Chaps. 1 and 2, we have provided an overview of spatial issues in public health, an introduction to typical analytical methods in epidemiology, and also an introduction to basic issues in geographical science. In Chap. 3, we unite notions of conceptual aspects of geographical information system and its usefulness in public health events. Spatial and temporal pattern of disease distribution has also been analyzed with suitable example. Chapter 4 describes the use of spatial statistics through exploration of methods, contests, and techniques associated with mapping disease data. In Chap. 5, we have provided an introduction to image processing techniques and methods for the analysis of spatial data and extend them to a particular issue of identifying disease cluster which is often needed in public health. In Chap. 6, we have analyzed the risk assessment of disease distribution in terms of public health. Finally, in Chap. 7, we have provided several issues and challenges of policy implementation undertaken to control the diseases using space technology. Throughout, we provide the case studies to illustrate the application of the methods which is well described in the text. Additional learning tools like maps, charts, figures, and tables have been provided throughout the text for better understanding.

This book provides a conceptual framework for the future researchers on geomedical application using remote sensing and GIS technology. The information in this book will be of immense significance for professionals, epidemiologists as well as to the amateur environmental scientists. This book directs and facilitates students of human geography to get a critical look at the theories and practices that jointly embrace GIS. We therefore hope that the book will be useful both as a standard reference and as a source of new research questions and hypotheses.

Midnapore, West Bengal, India

Gouri Sankar Bhunia
Pravat Kumar Shit

Acknowledgements

We are very much thankful to our teachers Dr. Dilip Kr. Pal, Dr. Nandini Chatterjee, Dr. Shreekant Kesari, Dr. Pradeep Das, Dr. Ajoy Mandal, Dr. Sunando Bandyopadhyay, Dr. Ramkrishna Maiti, Dr. Ashis Kumar Paul, Dr. Soumendu Chatterjee, Dr. Nilanjana Das Chatterjee, Dr. Utpal Roy, and Dr. Ratan Kumar Samanta for their lot of experiences, suggestions, encouragement and immense support throughout the work.

We would like to thank the anonymous reviewers who have acted as depended referees; their input was consistently constructive and has substantially improved the quality of the final product.

We would like to thank Dr. Jayasree Laha, Principal, Raja N. L. Khan Women's College, for her administrative support to carry on this project. We also acknowledge the contribution of the Department of Geography, Raja N. L. Khan Women's College, for providing logistic support and infrastructure facilities.

We would also like to thank Sanju Bera, Assistant Professor, Department of English, Sukumar Sengputa Mahavidyalaya, Keshpur, Paschim Medinipur, for his linguistic support.

We would like to thank our beloved parents for their infinite support and encouragement. We would also like to thank Ranita and Debjani, whose love, encouragement, and support kept us motivated up to the final shape of the book. Finally, the book has taken a number of years in its making and we therefore want to thank families and friends for their continued support.

We are also thankful to Springer Nature publishers, and Publishing Senior Editor Dr. Nabil Khélifi, Earth Sciences & MENA Program Springer—Heidelberg, Germany for their continuous support for publishing this book.

Midnapore, West Bengal, India

Gouri Sankar Bhunia
Pravat Kumar Shit

About This Book

Present book provides a research based study on vector borne disease in India through geospatial technology. The studies focused on the infectious disease i sub-tropical and ho humid environment. The present book also gathers creative research on geomedical applications using remote sensing and GIS technology. In this book, we have analyzed the basic concept and role of remote sensing, GIS and vector borne disease. Also, the present book represents the modern trends of geospatial technology in infectious disease risk assessment with appropriate illustration, statistical modelling and examples. This book comprises with spatial data, GIS, and spatial statistics to describe and interpret distributions of health related outcomes in public health problems.

Chapters 1 and 2, provides an overview of spatial issues in public health, an introduction to typical analytical methods in epidemiology, and an introduction to basic issues in geographical science. In Chap. 3, we have merged the ideas of geography and statistics through exploration of methods, challenges, and approaches associated with mapping disease data. In Chap. 4, have provide an introduction to image processing techniques and methods for the analysis of spatial data and extend them to a particular issue of identifying disease cluster which is often in interest in public health. Chapter 5 described about the ecological pattern and its associated with the vector borne disease pattern. In Chaps. 6 and 7 focused on the disease risk analysis and health-care planning policy. Finally, we have discussed several issues and challenges of policy implementation undertaken to control the diseases using space technology.

Contents

About the Authors

Gouri Sankar Bhunia received his Ph.D. from the University of Calcutta, India, in 2015. His Ph.D. dissertation work focused on environmental control measures of infectious disease (visceral leishmaniasis or kala-azar) using geospatial technology. His research interests include kala-azar disease transmission modeling, environmental modeling, risk assessment, data mining, and information retrieval using geospatial technology. He is Associate Editor and on the editorial boards of three international journal in Health GIS and Geosciences. He worked as a '*Resource Scientist*' in Bihar Remote Sensing Application Centre, Patna (Bihar, India). He is the recipient of the *Senior Research Fellow (SRF)* from Rajendra Memorial Research Institute of Medical Sciences (ICMR, India) and has contributed to multiple research programs kala-azar disease transmission modeling, development of customized GIS software for kala-azar 'risk' and 'non-risk' area, and entomological study.

Pravat Kumar Shit received his Ph.D. in Geography (Applied Geomorphology) from Vidyasagar University (India) in 2013, M.Sc. in Geography and Environment Management from Vidyasagar University in 2005, and PG Diploma in Remote Sensing & GIS from Sambalpur University in 2015. He is Assistant Professor in the Department of Geography, Raja N. L. Khan Women's College (Autonomus), Gope Palace, Midnapore, West Bengal, India. His main fields of research are soil erosion spatial modeling, badland geomorphology, gully morphology, water resources and natural resources mapping, and modeling and has published more than 45 international and national research articles in various renowned journals. His research work has been funded by the University Grants Commission (UGC), India. He is Associate Editor and on the editorial boards of three international journals in geography, and earth environment science.

Abbreviations

ABER	Annual Blood Examination Rate
AHP	Analytical Hierarchy Process
AIDS	Acquired Immune Deficiency Syndrome
ANN	Artificial Neural Network
API	Annual Parasite Index
AVHRR	Advanced Very-High-Resolution Radiometer
CR	Consistency Ratio
DTM	Digital Terrain Model
EID	Emerging Infectious Disease
GIS	Geographical Information System
GPI	Global Polynomial Interpolation
GPS	Global Positioning System
HIV	Human Immunodeficiency Virus
IDW	Inverse Distance Weighted
IRT	Inside Room Temperature
LPI	Local Polynomial Interpolation
LST	Land Surface Temperature
LULC	Land Use–Land Cover
ME	Mean Error
MODIS	Moderate Resolution Imaging Spectroradiometer
MSS	Multispectral Scanner System
MXL	Maximum Likelihood
NDVI	Normalized Difference Vegetation Index
NNA	Nearest Neighbour Analysis
NOAA	National Oceanic and Atmospheric Administration
NPP	Net Primary Productivity
NRDMS	National Resource Data Management System
NVBDCP	National Vector Borne Disease Control Programme
PCA	Principal Components Analysis
PF	*Plasmodium falciparum*

RBF	Radial Basis Function
RDVI	Re-normalized Difference Vegetation Index
RMSE	Root Mean Square Error
RS	Remote Sensing
SAM	Sandflies Abundance Mapping
SAVI	Soil-Adjusted Vegetation Index
SDE	Standard Deviation of Ellipse
SFR	Slide Falciparum Rate
SIMS	Summary Index of Malaria Surveillance
SPOT	France's Système Pour l'Observation de la Terre
SPR	Slide Positivity Rate
TIN	Triangulated Irregular Network
TM	Thematic Mapper
VL	Visceral Leishmaniasis
WGS	World Geodetic Survey
WHO	World Health Organization
WI	Wetness Index

List of Figures

List of Tables

Chapter 1
Introduction to Geoinformatics in Public Health

1.1 Introduction

Medical geography or health geography is a branch of human geography that focuses on the terrestrial aspect in the study of health prominence and the banquet of diseases. Additionally, it provides an idea of the location of individual health as well as its geographical distribution and its association with environmental factors. The concept of medical geography was first introduced by Hippocrates (5th–4th Century BCE). People have also been conscious of the development of disease dissemination through geographic regions for eras even during times when aetiology of infectious diseases was anonymous (e.g., the Black Death/plague, 1346–51 AD pandemic), which was conceded along trade paths from China to Europe. At present, medical geography has a lot of applications as well. Mapping plays an enormous role in this field. Maps are produced to demonstrate historic epidemics like the 1918 influenza or Google Flu Trends across the United States or Malaria, Leishmaniasis across the entire world. Medical geographers and public health professionals determine health strictly in terms of signs of illness such as morbidity and mortality. However, understanding of disease spreading may well be the most interesting and intriguing research area within the entire discipline of human geography.

Medical geography concerns about three main themes: disease ecology, health care delivery and environment and health. Disease ecology encompasses the investigation of infectious disease (e.g., malaria, filaria, leishmaniasis and HIV/AIDS) comprehending the geographical distributions of weather associated phenomenon, biotic and cultural portents interrelated with disease, along with the demographic, political and economic hurdles to assenting change. The research of health care provision embraces geographical measures of health care conveyance and patient activities and encompasses differences like discrepancies in health (health prominence and ease of contact), and de-institutionalization of the mentally ill. Environment and health is a comparatively novel emphasis for health

G. S. Bhunia and P. K. Shit, *Geospatial Analysis of Public Health*,
https://doi.org/10.1007/978-3-030-01680-7_1

geographers that appeals geography's long ritual in environmental hazards investigation along with health geography. Although the portraying of infection data can be comparatively straightforward, understanding geographically referenced disease data can occasionally be puzzling, mainly for non-infectious and chronic diseases (e.g., coronary heart disease and diabetes mellitus). Geographers have certain hindrances to be overawed to collect data. However, the leading problem is allied with the footage of a disease's location and the subsequent problem is connected with the precise identification of that disease.

1.2 Spatial Data for Public Health

Today's public health information is an embryonic field which emphasizes on the solicitation of information science and technology to public health rehearsal and investigation. Public health determinations have been based on the use and exploration of spatial data for several decades. Generally, public health varies from particular health because it exclusively depends on the health of people who are reluctant to expose it and the restriction of administrative framework. In 1854, Dr. Snow used a hand-drawn map to investigate the geographical location of London's cholera epidemic (Tuthill 2003). Snow assumed that cholera was spread through public water supplies, and determined the broad street pump as the outmost probable source of the cholera epidemic.

Data in a geographical contiguity is more prospectful to be predisposed by analogous factors and consequently pretentious in a similar manner. In 1890, Palm accomplished a study on geographical situation of rickets in an industrial urban area that had a cold and wet climate. Moreover, Florence Nightingale studies patient statistics and visualizes the reasons of mortality to establish that soldiers during Crimean war were suffering from disease connected to contaminations in hospitals circumstances and stimulate sanitary practices in medical amenities which consequently sustains millions of lives. Nevertheless, this historical remark can be of immense significance in demarcating patterns of the disease. Spatial analysis in public health not only pertains to geographical location of disease distribution but also to the structure and environmental conditions of a population.

1.3 Basic of Epidemiological Data

The examination of public health data usually comprises the concepts and tools of epidemiology demarcated by MacMahon and Pugh (1970) as the study of the dissemination and contributing factors of disease frequency. In most cases, the analyses of epidemiological data are based on annotations of disease incidence in a population of people "at risk". Normally, we want to narrate incidence patterns between groups of people suffering various levels of acquaintance to some factors having a putative influence on a person's risk of disease. Experimental studies endeavor to control all reasons that may adapt the connotations under study while observational studies cannot. Additionally, experimental investigations randomize consignment of the disputes of interest to investigational units to lessen the effect of any hysterical allied variable that may jiggle the relationship under study. In an observational study investigator detect issues of variable interest without conveying treatments to the subjects.

There are several ways by which the incidence of disease may be enumerated. The frequently used events of incidence and prevalence count both newly emergent and existing cases of the disease.

The general outcomes for epidemiologic investigations are as follows:

- *Mortality*: Mortality is the state of being mortal or liable to death; whereas the mortality rate governs the number of deaths in a particular population.
- *Illness*: Illness determines a disease is an exact abnormal state, a disorder of an erection or function that disturbs part or all of an organism. It can be determined through the physical signs, laboratory test etc.
- *Discomfort*: It is the sensation of infuriation, inflammation, or pain that, though not severe, is irritating. However, it reduces the capability to do normal activities.
- *Destitution*: Destitution is an unfortunate state in which a person is deficient in somewhat significant—like wealth, food, employment, companionship, or even hope.

1.4 Measures of Discase Frequency

Epidemiology is about recognizing associations between exposures and outcomes. To determine any association, the exposure and outcomes are first to be calculated in a quantitative approach. Then rates of occurrence are measured or calculated. These measures are referred "*measures of disease frequency*". Epidemiological measures of disease frequency are of five types:

- *Count*: Measures the number of population that meets the case definition. Calculating the extent of disease occurrence with a count is simple and helpful for definite purposes. It is more supportive to have a denominator under the count that indicates the size of the study population. For example, 20,000 cases of Kala-azar in Bihar in 2015.
- *Proportion*: Proportion determines the part of population affected by the disease. Proportions, also acknowledged as fractions are often stated as percentage that ranges from 0 to 1 or 0–100%. It can be calculated as:

$$A/(A + B)$$

where, A is population who meets the case definition.
B is the study population who does not meet the case definition and is at risk.

- *Ratio*: A ratio is simply one number divided by another. It is not dependent upon time. It is a measure of disease frequency. A ratio does not necessarily imply any particular relationship between the numerator (e.g., case definition) and the denominator (e.g. study population). For example, male-female ratio of Kala-azar disease in Bihar is 1:1.92.
- *Rate*: A rate is also one number divided by another, but the rate is reliant on time. It determines ratio over a certain period of time. An epidemiological rate will contain the following: disease frequency (numerator), unit of population size, and the time period during which the event occurred.

  For example: 14 cases per 1000 per year
  There are several ways to determine the disease rates, like:

– *Incidence rate* (*IR*)—Incidence rates calculate the occurrence of new cases of disease in a population. Conversely, incidence rates take into explanation the amount of the time that each individual persisted under surveillance and at risk of developing the outcome under study. It can be calculated as

$$IR = \frac{\textit{Number of new cases of disease during a specific}}{\textit{Population at risk during this time period}}$$

– *Prevalence rate* (*PR*)—It is directly related to the duration of disease. It can be determined as follows:

$$prevalence = incidence * average\,duration$$

Prevalence depends on the incidence rate (r) and the period of disease (T). Such as, if the incidence of a disease is low but the period of disease is lengthy, the prevalence will be high relative to the incidence. On the other hand, if the incidence of a disease is high and the period of the disease is short, the prevalence will be low relative to the incidence (Hennekens and Buring 1987). There are two collective estimation of prevalence rate:

(i) *Point prevalence*—Prevalence of a situation of interest at a exact time. It can vary from 0 to 100%. It can be measured from a cross sectional survey data by calculating the % with a particular disease on a particular date.

(ii) *Period prevalence*—Prevalence measured over an interval of time. It is the proportion of individual with a particular disease at any time during the interval. It can be calculated as

$$PR = \frac{All\,new\,\,and\,\,pre\text{-}existing\,cases\,during\,a\,given\,time\,period}{Population\,during\,the\,same\,time\,period} \times 10^n$$

- *Risk*: Risk is the proportion of individuals in a population (initially free of disease) who develop the disease within a specified time interval. Unwin defined risk is "*the probability that event will occur*".

  For example: 0.014 cases per person/year
  There are several ways to measure the risk, like:

- *Absolute risk* = incidence rate
- *Relative risk* = measure the strength of association between disease and without disease
- *Attributable risk*—Measure the ratio of disease in a population that can be attributed to the exposure.

The incidence risk assumes that the entire population at risk at the beginning of the study period has been followed for the specified time period for the development of the outcome under investigation.

1.5 Role of Remote Sensing in Public Health

Remote sensing (RS) refers to science and technologies that observe atmospheric and ground-based features from a distance. RS can identify features from remote-space, near-space, aerial, and terrestrial vantage points. Earth observation satellite allows us to quantify physical, chemical and biological factors (environmental occurrences and events) almost every place on the earth. Comprehensive supervision of the earth configuration, in both its natural and anthropological aspects, necessitates facts and figures that are timely, of known feature, durable and universal. RS provides such information and pays to refining our understanding of how the environment influences public health and welfare (Table 1.1 and Fig. 1.1). In medical geography, satellites such as Landsat's Multispectral Scanner (MSS) and Thematic Mapper (TM), the National Oceanic and Atmospheric Administration (NOAA)'s Advanced Very High Resolution Radiometer (AVHRR), and France's Système Pour l'Observation de la Terre (SPOT), can provide information about vegetation cover, landscape, structure, and water bodies in almost any region of the

Table 1.1 Use of remote sensing technology in some important vector borne disease application

Disease	Vector/reservoir	Location	Satellite/sensor	Methods/software	Remarks	Reference
Malaria	*An. albimanus*	Chiapas, Mexico	Aerial photos	Visual interpretation	Surrounding breeding sites of *An. albimanus* adult abundance were located at low elevations in flooded unmanaged pastures	Rodriguez et al. (1996)
	Anopheles albimanus	Tapachula, Chiapas, Mexico	LANDSAT (TM)	Multi-temporal satellite data	Using two remotely sensed landscape elements, the discriminant model was able to successfully distinguish between villages with high and low *An. albimanus*	Beck et al. (1997)
	Anopheles spp.	Gambia	NOAA (AVHRR), METEOSAT	Normalized difference vegetation index (NVDI) and Cold-Cloud Duration (CCD)	Processed to produce proxy ecological variables which have been extensively investigated for monitoring changes in the distribution and condition of different natural resources, including rainfall and vegetation	Thompson et al. (1996a, b)
	An. punctimaluca	Belize	SPOT (XS)	Discriminant function analysis, Canonical discriminant analysis	Habitat analysis and classification resulted in delineation of habitat types of mosquito defined by dominant life forms and hydrology	Rejmankova et al. (1998)
	An. subpictus	Lombok Island, Indonesia	JERS (optic)	Visible and near infrared radiometer to detect waterbodies, Overlay	Remote Sensing (RS), a Global Positioning System (GPS) and a Geographic Information System (GIS) were used to analyze relationship between *Anopheles subpictus* larval densities and environmental parameters	Anno et al. (2000)
	Anopheles spp.	Africa	LANDSAT (MSS, TM), SPOT, NOAA (AVHRR)		Investigating malaria epidemiology and assisting malaria control	Hay et al. (1998)
	Anopheles dints, Anopheles minimus	Assam, North-Eastern			Identified nature of the breeding ground for mosquitoes and their spreading patterns are not so complex as generally expected	Jeganathan et al. (2001)
	Anopheles spp.	Tanzania, Uganda, and Kenya/Africa	NOAA (AVHRR)	Discriminant analysis, multi-temporal meteorological satellite	The study identified land surface temperature as the best predictor of transmission intensity. Rainfall and moisture availability as inferred by Cold Cloud Duration (CCD) and the normalized difference vegetation index (NDVI), respectively, were identified as secondary predictors of transmission intensity	Omumbo et al. (2002)
		Kenya	RADARSAT 1	Land use/land cover analysis; texture analysis (eCognition software)	Object-oriented approach to image classification is taken in order to circumvent some of the limitations of traditional pixel-based classification of radar imagery	Kaya et al. (2002)

(continued)

Table 1.1 (continued)

Disease	Vector/ reservoir	Location	Satellite/sensor	Methods/software	Remarks	Reference
	Plasmodium vivax	Republic of Korea	IKONOS, Landsat	Supervised classification, cost comparison of chemoprophylaxis; PCI remote sensing software (PCI Geomatics, Richmond Hill, Ontario, Canada)	To determine whether an accurate estimate of the area covered by mosquito larval habitats	Masuoka et al. (2003)
	Aedes aegypti	Puntarenas, Costa Rica	ASTER, quickbird	Land cover analysis, artificial neural network (ANN); Idrisi Kilimanjaro software (J. R. Eastman, Clark University, Worcester, MA. 2004)	Developed for sampling specific mosquito larval habitats using GIS technology and high-resolution satellite imagery	Troyo et al. (2007)
	Anopheles gambiae	India	IRS WiFS	Land cover analysis, NDVI	Use of red and infra-red IRS WiFS multispectral data for land use/land cover mapping on 1:25,000 scale, and to map the malaria, and JE vector breeding habitats with spatial consistency	Palaniyandi (2014)
	P. falciparum	Mangalore, India	Landsat TM	Land use/land cover analysis; unsupervised classification, ENVI software v4.6 (ITT Visual Information Solutions, Boulder, CO, USA)	Detecting land cover changes and assesses their relationship with the burden of malaria	Mohan and Naumova (2014)
RRV (Ross River Virus)	*Culex annulirostris*	Brisbane, Australia	Colored aerial photos	MicroBRIAN image processing package	A rapid technique is being developed and assessed to identify urban breeding sites of *Culex annulirostris*	Dale and Morris (1996)
Lyme disease	*Ixodes scapularis*	Chappaqua and Armonk, Westchester	LANDSAT (TM)	Tasseled cap transformation index	A Geographic Information System (GIS) was used to spatially quantify and relate the remotely sensed landscape variables to Lyme disease risk category	Dister et al. (1997)
	Ixodes scapularis	Wisconsin, USA	NOAA (AVHRR)	NDVI, overlay	A Geographic Information System (GIS) was used to map distributions of human Lyme disease cases, ticks, and degree of vegetation cover	Kitron and Kazmierczak (1997)
	Ixodes scapularis	Wisconsin, Illinois, Michigan, USA	LANDSAT (TM)	Discriminant analysis, overlay	Environmental data were gathered at a local level (i.e., micro and meso levels), and a Geographic Information System (GIS) was used with several digitized coverages of environmental data to create a habitat profile	Guerra et al. (2002)
	Ixodes scapularis	Northeast to Southeast USA	NOAA (AVHRR)	Geostatistics, NDVI	Geostatistics (cokriging) was used to model the cross-correlated information between satellite-derived vegetation and climate variables and the distribution of the tick	Estrada-Peña (1998)

(continued)

Table 1.1 (continued)

Disease	Vector/reservoir	Location	Satellite/sensor	Methods/software	Remarks	Reference
Leishmaniasis	*Phlebotomus papatasi*	Saudi Arabia, Iran, Israel/Southeast Asia	NOAA (AVHRR)	NDVI	A computer model was developed using the occurrence of *P. papatasi* as the dependent variable and weather data as the independent variables	Cross et al. (1996)
	Lutzomyia spp.	Lagoinha, São Paulo, Brazil	LANDSAT (TM)	–	An area is characterized which may prove to be a macro-habitat for vectors, reservoirs and etiological agents	Miranda et al. (1998)
	Phlebotomus orientalis	Sudan/Africa	NOAA (AVHRR)	Logistic regression model, Normalized Difference Vegetation Index and land surface temperature	To estimate the probability of the presence of *P. orientalis* at each collecting site as a function of climatic and environmental variable	Thompson et al. (1999)
	Lutzomyia longipalpis	Teresina, Piauí, Brazil	LANDSAT (TM)	Spherical covariance structure, Spatial autocorrelation, NDVI (IDRISI software)	Demonstrate a method for modeling spatial autocorrelation within a mixed model framework, using data on environmental and socioeconomic determinants of the incidence of visceral leishmaniasis (VL) in the city of Teresina, Piauí, Brazil	Werneck and Maguire (2002)
	Phlebotomus argentipes	Vaishali district (Bihar, India)	IRS-1C LISS III	NDVI, supervised classification (ERDAS imagine v9.3)	Identify risk-prone areas of Kala-azar through GIS application tools	Sudhakar et al. (2006)
	Phlebotomus argentipes	Vaishali district (Bihar, India) and Lohardaga district (Jharkhand)	IRS-1C LISS III	NDVI, supervised classification (ERDAS imagine v9.3)	Identify the association between environmental factors and vector distribution	Paul et al. (2006)
	Phlebotomus argentipes	Northeastern Gangetic Plain,	NOAA (AVHRR)	Supervised classification (ERDAS imagine v9.3, ArcGIS v9.2)	The relationship between the incidence of VL and certain physio environmental factors was explored, using a combination of a geographical information system (GIS), satellite imagery, and data collected "on the ground"	Bhunia et al. (2010a, b)
	Phlebotomus argentipes	Vaishali district, Bihar, India	SRTM, Landsat 5 TM	DEM, NDVI (ERDAS imagine v9.3, ArcGIS v9.2)	To study the relationship between the incidence of Kala-azar and topography and vegetation density	Bhunia et al. (2010a, b)
	Phlebotomus argentipes	Vaishali district, Bihar, India	Landsat 5 TM	NDPI, nearest neighbour analysis, radial basic function interpolation (ERDAS imagine v9.3, ArcGIS v9.2)	Delineating the potential hydrological relationship between the vector and Kala-azar transmission, the associations between inland water bodies, sand fly prevalence, and Leishmania infections	Bhunia et al. (2011)

(continued)

Table 1.1 (continued)

Disease	Vector/ reservoir	Location	Satellite/sensor	Methods/software	Remarks	Reference
	Phlebotomus argentipes	Vaishali district in Bihar, India, and Lohardaga district in Jharkhand, India	IRS-LISS III, LISS IV	NDVI, supervised classification (maximum likelihood algorithm), spatial analysis, factor analysis (ERDAS imagine v9.3, ArcGIS v9.2)	Delineate the suitable habitats of the VL vector, *P. argentipes* density in relation to environmental characteristics between different ecosystems was assessed in endemic (Bihar) and non-endemic (Jharkhand) Indian states	Kesari et al. (2011)
	Phlebotomus argentipes	Muzaffarpur district (Bihar, India)	Landsat 5 TM	SAVI, WI, LST, and supervised classification (maximum likelihood algorithm) (ERDAS imagine v9.3, ArcGIS v9.2)	Examining relation with the environmental factors and vector distribution in a Kala-azar endemic region in Bihar, India	Bhunia et al. (2012a, b, c)
	Phlebotomus argentipes	Vaishali and Muzaffarpur districts (Bihar, India)	AVHRR, MODIS, Landsat TM, LISS IV	Thematic maps and satellite supervised classification (ERDAS imagine v9.3, ArcGIS v9.2)	Examining the relationship between LULC classes and their suitability for vector habitats in areas endemic for Kala-azar at different spatial scale	Bhunia et al. (2012a, b, c)
Flariasis	*Culex pipiens*	Nile River Delta/ Africa	NOAA (AVHRR)	TeraScan software to determine dTs, Overlay method	Correlation between Bancroftian filariasis distribution and diurnal temperature differences in the southern Nile delta	Thompson et al. (1996a, b)
Trypanosomiasis	*Glossina spp.*	Kenya/Africa	LANDSAT (TM)	Spectral bands of Landsat TM, multiple regression	Satellite imagery, Geographic Information Systems (GIS) and spatial statistics provide tools for studies of population dynamics of disease vectors in association with habitat features on multiple spatial scales	Kitron et al. (1996)
	Glossina spp.	Côte d'Ivoire and Burkina Faso/Africa	NOAA (AVHRR)	Temporal Fourier-processed surrogates for vegetation, temperature and rainfall derived from meteorological satellites	The application of remotely-sensed, satellite data to the problems of predicting the distribution and abundance of tsetse flies in West Africa	Rogers et al. (1996)
	Glossina spp.	Southern Africa	NOAA (AVHRR)	Linear discriminant analysis, maximum likelihood classification	Study about the distribution of *Glossina morsitans* centralis, *Glossina morsitans* and *Glossina* pallidipes using a range of multivariate techniques applied to climate and remotely sensed vegetation data	Robinson et al. (1997)
	Glossina tachinoides	Togo/Africa	NOAA (AVHRR), METEOSAT	0.125° raster or grid-based Geographic Information System data used	Addresses the problem of generating tsetse distribution and abundance maps from remotely sensed data, using a restricted amount of field data	Hendrickx et al. (1999)
	Glossina spp.	Africa	NOAA (AVHRR)	Multi-temporal satellite data; temporal fourier analysis; biological, process-based models	Descriptions of the different components of transmission, from the parasites to the affected hosts, eventually developed to include geographical dimensions	Rogers (2000)
	Glossina spp.	Burkina Faso/Africa	LANDSAT (TM) SPOT			De La Rocque et al. (2001)

Source de Moraes Correia et al. (2004), Cad. Saúde Pública, Rio de Janeiro, 20(4):895–896; Bhunia et al. (2013), ISRN infectious disease

NDVI normalized difference vegetation index, *WI* wetness index, *SAVI* soil adjusted vegetation index, *LST* land surface temperature, *NDPI* normalized difference pond index, *IRS* Indian remote sensing system, *LISS* linear imaging self-scanning, *TM* thematic mapper, *AVHRR* advanced very high resolution radiometer, *MODIS* moderate resolution imaging spectrora diometer

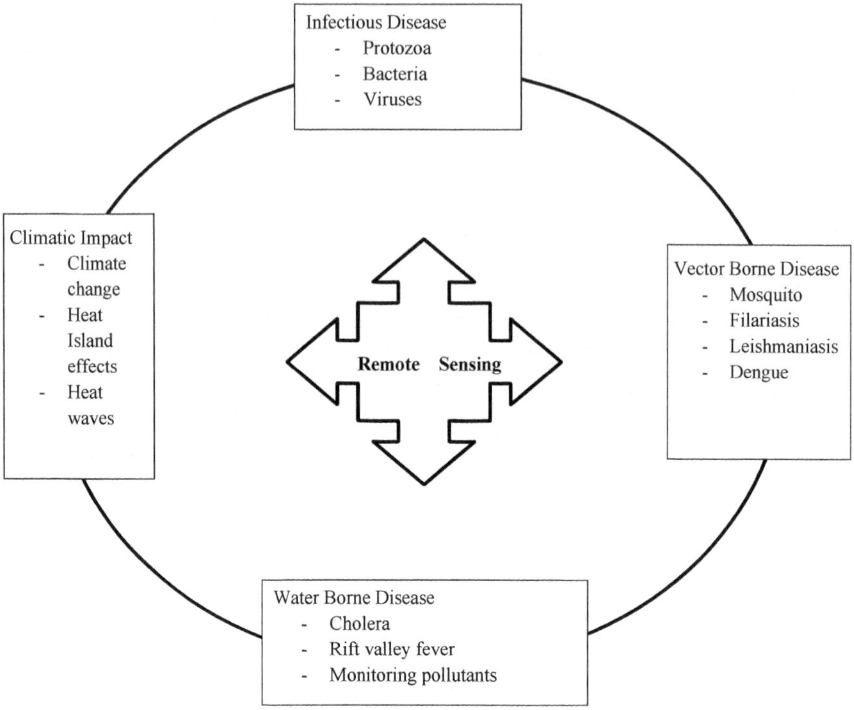

Fig. 1.1 Schematic diagram of remote sensing application in public health

globe—information that can be extremely valuable in health research that examines environmental factors in disease dissemination (Beck et al. 2000).

For example, when considering the association between climate and vector borne diseases, the succeeding associations among the distribution and life cycle of the vector, outbreaks of the disease, the impact on endemicity and the socio-economic drivers of the diseases should be measured. To generate a climatological record that can be likened to the above components of a disease, remote sensing and GIS is extremely relied upon. Current enhancement in spatial and temporal resolution of climatic variables has permitted for more healthy investigations of the connection of climate and diseases.

Remote Sensing has been used to predict cholera outbreaks in Bangladesh that is based on large-scale oceanic algal blooms (Ali et al. 2002); to identify snail habitat for schistosomiasis control in China (Guo-Jing et al. 2002); to predict the distribution of urinary schistosomiasis in Tanzania using land surface temperature (LST) and the normalized difference vegetation index (NDVI) (Brooker 2002); to map the distribution of intestinal schistosomiasis in Uganda using AVHRR (Kabatereine et al. 2004); to assemble a household-GIS database for health lessons in Karachi, Pakistan, by high resolution IKONOS imagery where GPS receivers failed because of structural barriers such as tall buildings (Ali et al. 2004); to quantify areas of reduced risk of hantavirus pulmonary syndrome in the United States using Landsat Thematic Mapper imagery (Glass et al. 2000); to envisage intestinal schistosomiasis infection in school children in the Côte d'Ivoire (Raso et al. 2005); to risk map VL in Sudan using NDVI and climate data (Elnaiem et al. 2003); to anticipate malaria epidemics in sub-Saharan Africa using climatic variables to predict vector habitat (Rogers et al. 2002); to determine small-area clustering of malaria in Nandi District, Kenya via land cover types recovered from the Digital Landsat Enhanced Thematic Mapper+ (ETM+) (Brooker et al. 2004); and to identify environmental factors that could predict Ascaris infections in South Africa (Saathoff et al. 2005).

The use of Remote Sensing techniques to map vector distribution and disease risk has evolved considerably during the last two decades. The complexity of methods range from using simple correlations between spectral signatures from different land use/land cover types and species abundance (Sithiprasasna et al. 2005) to complex techniques that link satellite-derived seasonal environmental variables to vector biology (Rogers et al. 2002). An assessment of different modeling approaches for mapping vector and vector-borne diseases is discussed by Rogers (2006). An outline of the accessibility of environmental satellite data for mapping infectious diseases can be found in Hay et al. (2006).

Role of Remote Sensing data in disease epidemiology involves retrieving environmental variables that characterize the vector ecosystem such as land cover, temperature, humidity or vapor pressure, and precipitation. However, measuring meteorological and climate variables near the surface is more tricky, and repeatedly, empirical methods were employed (Rogers 1991). The Normalized Difference Vegetation Index (NDVI), which exploits the strong contrast in the reflectance of vegetation in the red and near infrared wavelengths, is a commonly used index to characterize vegetation dynamics (Townshend and Justice 1986). An amalgamation of vegetation indices, surface reflectance, and temperature measurements have been used by epidemiologists to model vector ecosystems (Rogers et al. 2002).

Multispectral, microwave or thermal satellite imagery cannot be employed to monitor sand fly or vectors in a straight line from space, but can be used to recognize the favourable environment or breeding places. Remote Sensing technologies, which allow the mapping of environmental variables, have already been used in different epidemiological studies (Bhunia et al. 2010a, b), but so far only rarely deal with vector borne disease. Few studies are available that include the extraction of environmental indicators like meteorology, vegetation and altitude etc. (Sudhakar et al. 2006). Nieto et al. (2006) developed an ecological niche model to delineate the distribution and potential risk zone of Visceral leishmaniasis (VL).

1.6 Geographic Information Systems (GIS) in Public Health Research

The application of geographic information systems (GIS) to public health exercise has prodigious prospective for enlightening our understanding of the ecology and reasons of complex health problems, and for managing the policy and appraisal of effective population based programs and policies. According to Bill (1999) a Geographic Information System (GIS) is a computer-supported system consisting of hardware, software, data and the consequent applications. By means of GIS, data can be digitally recorded and edited, stored and reorganized, shaped and analyzed as well as presented in an alphanumerical and graphic mode. In its definition the (World Health Organization (WHO) 1999) states another essential: the trained staff. The spatial dimension of health and health care has been being noted since ancient times (Picheral 1994). There are three different types of geographic-epidemiological studies: disease mapping, ecological studies and migrant studies (English 1992). The

Table 1.2 Important analyzing methods of Geographical Information System (GIS)

Method	Description	Reference
Data base query	Identification of objects on the basis of user-defined selection terms	Huang et al. (2012)
Geometrical calculations	All functions carrying out calculations on the basis of geometry: distances, longitudes, areas, angles, differences in altitude etc.	Keegan and Dushoff (2014)
Overlay, clip, merge	Using these techniques new variables can be calculated, for instance, to check which measurement points are within a certain area	Achu (2008)
Buffering	Construction of zones (buffer) of determined dimension by points, lines and areas	Palaniyandi (2012)
Density estimation	Estimation of spatial density of geometric objects on the basis of user-defined conditions (e.g. Kernel estimation)	Hollingsworth et al. (2015)
Interpolation	Estimation of missing data on the basis of space-related relations and distribution of known data (e.g, Kriging); smoothing methods; construction of smoothed (generalized) patterns of attribute data (surfaces) (e.g., surface trend analysis)	Diuk-Wasser et al. (2010)
Analysis of space-related distribution	Check of space-related data in view of correlation and cluster by using visualization methods and geo-statistical methods (auto-correlation, Moran's coefficient, Nearest Neighbor Procedure etc.)	Ratmanov et al. (2013)
Modeling and simulation	Development of models and scenarios on the basis of geometric and attribute data, in particular tempo-spatial distribution- and spreading models etc.	Shah and Gupta (2013)

functionalities of GIS include the following selected aspects (Table 1.2) that are provided by the different scientists and researchers' worldwide (Clarke et al. 1996).

There are noteworthy methodological concerns which must be addressed so as to confirm that map yields are interpretable and not ambiguous. GIS can abridge vast extents of tabular data into convincing visual maps that can offer prevailing intuitions and engross the attention of policy makers and the public. GIS has been utilized to map the national distribution of lymphatic filariasis in Nepal (Sherchand et al. 2003); to invent threat maps of lymphatic filariasis in Africa based on climactic variation (Lindsay and Thomas 2000); to unmask the profound heterogeneity of malaria risk in magisterial districts of South Africa (Booman et al. 2000); to predict the spatial distribution of Schistosomiasis in Tanzania for use in a national mass drug treatment control program (Clements et al. 2006); to model patterns of African Trypanosomiasis in southern Cameroon (Muller et al. 2004); to expand models integrating livestock biomass, tsetse flies, farming systems, clinical disease, and land use for the control of African Trypanosomiasis (Hendricks et al. 2001); to establish the spatial outline of African Trypanosomiasis in Cote D'Ivoire using GPS and ground-collected information on households, agriculture, and vegetation (Courtin et al. 2005); to envisage community predominance of Onchocerciasis in the Amazon (Carabin et al. 2003); to map the global distribution of trachoma and Trichiasis (Polack et al. 2005); to identify areas of high risk for Giardiasis in Canada (Odoi et al. 2003); to determine the spatial distribution of visceral leishmaniasis infection in Africa and India (Bhunia et al. 2010a, b); and to construct a disease atlas of helminth infection in sub-Saharan Africa (Brooker et al. 2000).

The potential of GIS has yet to be revealed in at least two areas: a thematic one (i.e., policy making) and, suitably enough, a geographical under-developed region (Table 1.3). GIS applications related to Kala-azar have been introduced and used in, for example, the surveillance and monitoring of diseases (Kalluri et al. 2007), in environmental health (Bhunia et al. 2010a, b), quantifying environmental hazards and their influence on public health (Salomon et al. 2006), and for policy and planning purposes (Clements et al. 2006). In India, for example, GIS systems have been used in vector control research (Bhunia et al. 2012a, b, c) for studying and mapping of non-communicable diseases (Raban et al. 2009). The application of GIS in a public health circumstance can be a resource intensive activity, demanding a substantial speculation.

1.7 Statistical Methods for Spatial Data in Public Health Research

Statistical data analysis is now the most consistent and recognized set of tools to evaluate spatial datasets. Yet the solicitation of statistical practices of spatial data appearances is an imperative challenge, as conveyed in Tobler's (1969) first law of

Table 1.3 Use of Geographic Information System (GIS) in some important vector borne disease application

Disease	Location	Methods/software	Remarks	Reference
Malaria	Sub-Saharan Africa	Orthograph, aerial photograph, MapInfo software (version 4, MapInfo Corporation, New York, USA); Bentley and Intergraph software products (Symmetry Systems Inc., New York, USA) and Global Positioning System (Optron Precise Positioning Solutions, Johannesburg, South Africa)	Provides an example of how a geographical information system can contribute to the planning of malaria control programmes	Booman et al. (2000)
	South Africa	Geographical Information System based Malaria Information System	To process data timeously into a usable format is discussed, as well as its relevance to malaria research, appropriate malaria control measures, tourism, and social and economic development	Martin et al. (2002)
	Indonesia	Literature survey	Discussion of strategies that can be used to overcome some of problems like technological problems accurate data on the disease and how it is reported; basic environmental data and demographic data	Sipe and Dale (2003)
	Sub-Saharan Africa	Overlay and spatial analysis (ArcView version 3.2)	To relate stability of malaria transmission to biologic characteristics of vector mosquitoes throughout the world	Kiszewski et al. (2004)
	Valle del Cauca, Colombia	Malaria Climatic Convenience Index (MCCI), Malaria Natural Convenience Index (MNCI), Malaria Risk Transmission Index (MRTI)	To develop a methodology for mapping malaria risk, which integrates physical variables such as temperature, precipitation and geomorphologic features with related aspect to human being, which in this study will be recognized as anthropic variables	Rincón-Romero and Londoño (2009)

(continued)

Table 1.3 (continued)

Disease	Location	Methods/software	Remarks	Reference
	Orissa, India	Overlay, index model Arc/View	Identifies the risk factors associated with high malaria transmission and focused intervention based on geomorphological parameters, land use, soil type, water bodies and drainage network	Daash et al. (2009)
	Laos	NAVSTAR satellite system; KASHMIR 3D Version 8.0.9 Beta; Mandara for Windows Version 9.10	Geographic Information System (GIS) maps were developed using the data collected in an active case detection survey	Shirayama et al. (2009)
	Developing countries	Literature survey	Focuses on how advances in mapping, Geographic Information System, and Decision Support System technologies, and progress in spatial and space-time modeling, can be harnessed to prevent and control these diseases	Eisen and Eisen (2010)
	Bangladesh	PubMed database	Recent progress of malaria mapping in Bangladesh with GIS, GPS, and RS, and identified potential future applications and contributions of geospatial technologies to eliminate malaria in the country	Kirk et al. (2015)
RRV (Ross River Virus)	Leschenault estuary, WA, (south-west Australia)	Buffer zone, Spatial analysis	Investigate the relationship between risk of Ross River virus (RRV) infection and proximity to mosquito-breeding habitat surrounding a tidal wetland ecosystem	Vally et al. (2012)
	Australia	Principal Component Factor analysis, K-means cluster analysis; Spatial distribution analysis	To assess the relationship between socio-environmental variability and the transmission of RRV using spatio-temporal analysis	Hu et al. (2005)
	Australia	Spatial analysis; Principal Component Factor analysis and regression, chi-square tests	Spatial distribution was investigated using census data at the suburb level	Muhar et al. (2000)

(continued)

Table 1.3 (continued)

Disease	Location	Methods/software	Remarks	Reference
Lyme disease	USA	Risk model, logistic regression analysis	Identify and locate residential environmental risk factors for Lyme disease	Glass et al. (1995)
	United States	Risk model and spatial analysis, ARC/INFO and ArcView GIS (ESRI, Redlands, CA), Trimble Geoexplorer (Trimble Navigation, Ltd., Sunnyvale, CA)	To determine the distribution of *I. scapularis* in the upper Midwest based on data from Wisconsin, northern Illinois, and the Upper Peninsula of Michigan, and to explain the environmental factors that facilitate or inhibit the establishment of *I. scapularis*	Guerra et al. (2002)
	Ontario	Multi-Criterial Decision Making Model, Spatial analysis	Spatial distribution of endemic tick populations at the dissemination area (DA) level and the potential role of white-tailed deer in the spatial expansion of Lyme ticks; and determine the relationship between the tick *B. burgdorferi* bacterium and deer establishment	Chen et al. (2015)
Leishmaniasis	Vaishali district (Bihar, India)	Inverse distance weightage, Geostatistics, hotspot, spatial autocorrelation (ArcGIS software v 9.0)	Investigated the spatio-temporal patterns and hotspot detection for reporting Kala-azar cases in Vaishali district based on spatial statistical analysis	Bhunia et al. (2013)
	Muzaffarpur district (Bihar, India)	Spatial analysis and standard deviation of ellipse, spatial statistics (ArcGIS v9.2)	Examining disease distribution in a Kala-azar endemic region in Bihar, India	Bhunia et al. (2012a, b, c)
	Muzaffarpur district (Bihar, India)	Database queries, spatial analysis (ArcGIS v9.2)	The spatial distribution of reported Kala-azar cases in the 4 study periods of Muzaffarpur district, Bihar, India	Malaviya et al. (2011)

(continued)

Table 1.3 (continued)

Disease	Location	Methods/software	Remarks	Reference
	Venda Nova, Belo Horizonte, Minas Gerais, Brazil	Literature based study	The use of an automated database allied with geoprocessing tools may favor control measures of VL, especially with regard to the evaluation of control actions carried out	Saraiva et al. (2011)
	Iran	Global clustering methods including the average nearest-neighbour distance, Moran's I, general G indices and Ripley's K-function	To analyse yearly spatial distribution and the possible spatial and spatio-temporal clusters of the disease to better understand spatio-temporal epidemiological aspects of ZCL in rural areas of an endemic province	Mollalo et al. (2014)
	Brazil	Digital database, spatial analysis, overlay	Produce distributional maps of the phlebotomine vectors of American cutaneous leishmaniasis and superimpose these data with American cutaneous leishmaniasis disease records for the historical periods	Shimabukuro et al. (2010)
Filariasis	Nigeria	Spatial analysis of geographically referenced data	Focuses on how the use of Geographical Information System (GIS) can be harnessed for surveillance, prevention and control of LF and malaria	Okorie et al. (2014)
	Sri Lanka	Digital database and spatial analysis	To develop a site directed Geographic Information System (GIS) map of Lymphatic Filariasis (LF)	Wijegunawardana et al. (2012)
	Andhra Pradesh, India	Spatial database generation, GPS (ArcGIS Engine-9.2)	To present a spatial mapping and analysis of filariasis data over the historical period	Upadhyayula et al. (2012)

(continued)

Table 1.3 (continued)

Disease	Location	Methods/software	Remarks	Reference
Trypanosomiasis	Kenya	Database generation and spatial analysis, Global Positioning System (GPS), MapInfo Software	To map the spatial and temporal distribution of SS and determine possible risk factors associated with the disease	Rutto and Karuga (2009)
	Zambia	Spatial analysis, Buffering, multivariate (maximum likelihood) analysis	To identify areas where intervention is most likely to be technically, economically, socially and environmentally sustainable	Robinson et al. (2010)
	Africa	Georeferencing, Database generation	Mapping the distribution of human African trypanosomiasis in time and space	Simarro et al. (2010)
	Zambia	Decision-tree approach combined with a multiple-criteria evaluation (MCE)	To show how remotely sensed and other environmental data can be combined in a decision support system to help inform tsetse control programmes in a manner that could be used to limit possible detrimental effects of tsetse control	Symeonakis et al. (2007)

geography: "everything is related to everything else, but near things are more related than distant things". Statistical analysis which covenants with geographically referenced data is designated as the science of spatial statistics. Standard statistical approaches undertake independence of observation. When employing this method to examine spatially interrelated data, the standard error of the covariate parameters is undervalued so the statistical significance is overemphasized. Spatial statistical procedures integrate spatial association along with the way of geographical contiguity is defined. Proximity further is governed by the geographical information that can be obtainable at areal/regional level or at point location level.

- Areal unit data are gathered over adjoining units (countries, state, districts, and survey zones) which divide the entire study region. Proximity in span is demarcated by their adjacent structure.
- Point referenced data are composed at stationary locations (household, villages) over an incessant study region.

However the contiguity in spatial statistical data is determined by the remoteness between sample locations. The crucial part of probabilities stimulates the practice of statistical methods to examine public health data and the usage of geostatistical approaches to

- Apprise various rate perceived from various geographical areas,
- Discrete arrangement from noise,
- Recognize disease clusters, and,
- Evaluate the connotation of latent exposures.

Furthermore, these methods permit to enumerate ambiguity in our assessments, forecasts, and maps and make available the practicalities for statistical inferences with spatial data.

GIS software and tools can be employed to generate covariates for inclusion in statistical model and to envisage the output from statistical models. Spatial statistics deliver regions modeling and extrapolations method for portraying interfaces from geographically referenced data. Geostatistics offers body of approaches for spatial smoothing and for accenting for spatial covariates in appearing spatial surface.

GIS based statistical models are employed to assess vector incidence or profusion within a specific geographic area. Basic spatial modeling methods comprise—interpolation based on spatial dependence in vector and extrapolation based on relations between vector data and environmental or socio-economic predictor variables. For instance, zones with high vector profusion or maximum disease occurrence often occupy on other areas with high vector abundance or high disease incidence, and the resemblance in their influencing variable losses with growing space. In these cases, kriging or supplementary categories of interpolation models are employed to create smooth interpolated maps of the influencing factors (Bunnell et al. 2003; Diuk-Wasser et al. 2010). GIS based extrapolation model, software is first employed to excerpt geographically categorical data for environmental variables of interest for the point locations or physical areas where the data were composed. Afterward, a prognostic model is established in a statistical software package and the model calculation is then useful in the GIS. For example, using the Raster Calculator in ArcGIS, continuous spatial surfaces were developed that present a estimated risk of exposure to vectors or vector borne pathogen (Craig et al. 2008; Honório et al. 2009).

1.8 Global Positioning System (GPS) in Public Health Research

Hand-held GPS is a technology developed by the United States Department of Defense that uses a constellation of 24–32 medium earth orbiting satellites to pinpoint a user's location, speed, direction, and time (King et al. 2004; Strom 2002). Re-developed for civilian use under the issue of Ronald Reagan in 1983, GPS today is utilized in a variety of geospatial applications, from superior computer cartography to aboard consumer automobile navigation systems (Pellerin 2006). Mention may be made of Tran et al. (2008)'s use of GPS to identify and map larval and adult populations of *Anopheles hyrcanus* to examine the potential of re-emergence of malaria in Southern France; Zeilhofer et al. (2007)'s identification of habitat suitability of *Anopheles darlingi*, a vector of malaria, with GPS around hydroelectric plants in Mato Grosso State, Brasil; and Dwolatzky et al. (2006)'s accomplishment of GPS into a personal digital assistant (PDA) for health care workers to locate remote home sites of tuberculosis cases in support of a tuberculosis control programme in South Africa.

1.9 Conclusion

At present, medical geography has a number of uses as well. Technological progresses continue with medical improvements. Meanwhile, the geographical dissemination of the disease is still a large substance of significance however; mapping plays an enormous role in the field. Although, geographers have some difficulties to be overawed when collecting data, the prime problem is allied with recording a disease's location. The earlier report suggested that social disparities and environmental factors are more lean towards key determinants of discriminations in health than access to health care. With growing interest in health GIS, the epidemiological method, assumed in the field of geography of disease relied increasingly on the statistical modeling of the geographical dissemination of diseases and their distribution in time and space. These methods allow health related information to be exhibited, and enable the visualization and monitoring of infectious disease. Uses of Remote Sensing and Geographic Information Systems are quickly gaining recognition as effective means to answer complex, ecological questions in health endorsement, public health, medicine, and epidemiology (Miranda and Dolinoy 2005; Foody 2006). The optimal use of RS and GIS will require not only continued innovation in technology and application but also something that is not yet visible: a continuous flow of information between disciplines and across borders, focusing on the end of the result.

References

Achu DF (2008) Application of gis in temporal and spatial analyses of dengue fever outbreak: case of Rio De Janeiro, Brazil. Linköpings Universitet Linköping, Sweden. Available at: https://www.diva-portal.org/smash/get/diva2:210116/FULLTEXT01.pdf

Ali M, Emch M, Donnay JP, Yunus M, Sack RB (2002) Identifying environmental risk factors for endemic cholera: a raster GIS approach. Health & Place 8(3):201–210

Ali M, Rasool S, Park JK, Saeed S, Ochiai R, Nizami Q, Acosta CJ, Bhutta Z (2004) Use of satellite imagery in constructing a household GIS database for health studies in Karachi, Pakistan. Int J Health Geogr 3(1):20

Anno S, Takagi M, Tsuda Y, Yotopranoto S, Dachlan YP, Bendryman SS et al (2000) Analysis of relationship between Anopheles subpictus larval densities and environmental parameters using Remote Sensing (RS), Global Positioning Systems (GPS) and a Geographic Information System (GIS). Kobe J Med Sci 46:231–243

Aparício C, Dantas-Bittencourt M (2003) Análise especial da leishmaniose tegumentar americana. In: Anais do XI Simpósio Brasileiro de Sensoriamento Remoto; Abr 5–10; Minas Gerais, Brasil. Belo Horizonte: Instituto Brasileiro de Pesquisas Espaciais

Beck LR, Lobitz BM, Wood BL (2000) Remote sensing and human health: new sensors and new opportunities. Emerg Infect Dis 6(3):217–227

Beck LR, Rodriguez MH, Dister SW, Rodriguez AD, Washino RK, Roberts DR (1997) Assessment of a remote sensing based model for predicting malaria transmission risk in villages of Chiapas, Mexico. Am J Trop Med Hyg 56:99–106

Bhunia GS, Chatterjee N, Kumar V, Siddiqui NA, Mandal R, Das P (2012a) Delimitation of Kala-azar risk areas in the district of Vaishali in Bihar (India) using a geo-environmental approach. Memórias do Instituto Oswaldo Cruz 107(5):609–620

Bhunia GS, Kesari S, Chatterjee N, Kumar V, Das P (2012b) Localization of Kala-azar in the endemic region of Bihar, India based on land use/land cover assessment at different scales. Geospatial Health 6(2):177–193

Bhunia GS, Kesari S, Chatterjee N, Kumar V, Das P (2012c) Seasonal relationship between normalized difference vegetation index and abundance of the Kala-azar vector in an endemic focus in Bihar, India. Geospatial Health 7(1):51–62

Bhunia GS, Kesari S, Chatterjee N, Kumar V, Das P (2013) The burden of visceral leishmaniasis in India: challenges in using remote sensing and GIS to understand and control. ISRN Infect Dis 1–14

Bhunia GS, Kesari S, Chatterjee N, Pal DK, Kumar V, Ranjan A, Das P (2011) Incidence of visceral leishmaniasis in the Vaishali district of Bihar, India: spatial patterns and role of inland water bodies. Geospatial Health 5(2):205–215

Bhunia GS, Kesari S, Jeyaram A, Kumar V, Das P (2010a) Influence of topography on the endemicity of Kala-azar: a study based on remote sensing and geographical information system. Geospatial Health 4(2):155–165

Bhunia GS, Kumar V, Kumar AJ, Das P, Kesari S (2010b) The use of remote sensing in the identification of the eco-environmental factors associated with the risk of human visceral leishmaniasis (Kala-azar) on the Gangetic plain, in north-eastern India. Ann Trop Med Parasitol 104(1):35–53

Bill R (1999) Grundlagen der Geo-Informationssysteme, 4th ed. Band 1 (Hardware, Software und Daten). 2. Aufl. Wichmann Verlag, Heidelberg pp 454

Booman M, Durrheim DN, La Grange K, Martin C, Mabuza AM, Zitha A, Mbokazi FM, Fraser C, Sharp BL (2000) Using a geographical information system to plan a malaria control programme in South Africa. Bull World Health Organ 78(12):1438–1444

Brooker S (2002) Schistosomes, snails, and satellites. Acta Trop 82:207–214

Brooker S, Clarke S, Njagi JK, Polack S, Mugo B, Estambale B, Muchiri E, Magnussen P, Cox J (2004) Spatial clustering of malaria and associated risk factors during an epidemic in a highland area of western Kenya. Trop Med Int Health 9(7):757–766

Brooker S, Rowlands M, Haller L, Savioli L, Bundy DAP (2000) Towards an atlas of human helminth infection in sub-Saharan Africa: the use of geographic information systems (GIS). Parasitol Today 16(7):303–307

Bunnell JE, Price SD, Das A, Shields TM, Glass GE (2003) Geographic information systems and spatial analysis of adult Ixodes scapularis (Acari: Ixodidae) in the Middle Atlantic Region of the USA. J Med Entomol 40:570–576

Carabin H, Escalona M, Marshall C, Vivas-Martinez SV, Botto C, Joseph L, Basáñez M-G (2003) Prediction of community prevalenceof human onchocerciasis in the Amazonian onchocerciasis focus: Bayesian approach. Bull World Health Organ 81:482–490

Chen D, Wong H, Belanger P, Moore K, Peterson M, Cunningham J (2015) Analyzing the correlation between deer habitat and the component of the risk for lyme disease in Eastern Ontario, Canada: a GIS-based approach. ISPRS Int J Geo-Inf 4:105–123

Clarke K, McLafferty S, Tempalski B (1996) On epidemiology and geographic information systems: a review and discussion of future directions. Emerg Infect Dis 2:85–92

Clements ACA, Lwambo NJS, Blair L, Nyandindi U, Kaatano G, Kinung'hi S, Webster JP, Fenwick A, Brooker S (2006) Bayesian spatial analysis and disease mapping: tools to enhance planning and implementation of a schistosomiasis control programme in Tanzania. Trop Med Int Health 11(4):490–503

Connor SJ, Thompson MC, Flasse SP, Perryman AH (1998) Environmental information systems in Malaria: risk mapping and epidemic forecasting. Disasters 22:39–56

Courtin F, Jamonneau V, Oké E, Coulibaly B, Oswald Y, Dupont S, Cuny G, Doumenge J-P, Solano P (2005) Towards understanding the presence/absence of human African trypanosomosis in a focus of Côte d'Ivoire: a spatial analysis of the pathogenic system. Int J Health Geogr 4(1):27

Craig W, Tepfer M, Degrassi G, Ripandelli D (2008) An overview of general features of risk assessments of genetically modified crops. Euphytica 164:853–880. https://doi.org/10.1007/s10681-007-9643-8

Cross ER, Newcomb WW, Tucker CJ (1996) Use of weather data and remote sensing to predict the geographic and seasonal distribution of Phlebotomus papatasi in Southwest Asia. Am J Trop Med Hyg 54:330–332

Daash A, Srivastava A, Nagpal BN, Saxena R, Gupta SK (2009) Geographical information system (GIS) in decision support to control malaria—a case study of Koraput district in Orissa, India. J Vector Borne Dis 46(1):72–74

Dale PER, Morris CD (1996) Culex annulirostris breeding sites in urban areas: using remote sensing and digital image analysis to develop a rapid predictor of potential breeding areas. J Am Mosq Control Assoc 12:316–320

De La Rocque S, Michael JF, De Wispelaere G, Cuisance D (2001) New tools for the study of animal trypanosomiasis in the Sudan: model-building of dangerous epidemiological passage by remote sensing geographic information systems. Parasite 8:171–195

Dister SW, Fish D, Bros SM, Frank DH, Wood BL (1997) Landscape characterization of peridomestic risk for Lyme disease using satellite imagery. Am J Trop Med Hyg 57:687–692

Diuk-Wasser MA, Vourc'h G, Cislo P, Hoen AG, Melton F, Hamer SA, Rowland M, Cortinas R, Hickling GJ, Tsao JI (2010) Field and climate-based model for predicting the density of host-seeking nymphal Ixodes scapularis, an important vector of tick-borne disease agents in the eastern United States. Global Ecol Biogeogr 19:504–514

Dwolatzky B, Trengove E, Struthers H, McIntyre J, Martinson N (2006) Linking the global positioning system (GPS) to a personal digital assistant (PDA) to support tuberculosis control in South Africa: a pilot study. Int J Health Geogr 5(1):34

Eisen L, Eisen RJ (2010) Using Geographic Information Systems and decision support systems for the prediction, prevention, and control of vector-borne diseases. Ann Rev Entomol 56:41–61

Elnaiem DA, Schorscher J, Bendall A, Obsomer V, Osman ME, Mekkawi AM, Connor SJ, Ashford RW, Thomson MC (2003) Risk mapping of visceral leishmaniasis: the role of local variation in rainfall and altitude on the presence and incidence of Kala-azar in eastern Sudan. Am J Trop Med Hyg 68(1):10–17

English D (1992) Geographical epidemiology and ecological studies. In: Elliot P, Cuzick J, English D, Stern R (eds) Geographical and environmental epidemiology: methods for small-area studies. Oxford Press, Oxford, pp 3–13

Estrada-Peña A (1998) Geostatistics and remote sensing as predictive tools of tick distribution: a cokriging system to estimate Ixodes scapularis (Acari: Ixodidae) habitat suitability in United States and Canada from advanced very high resolution radiometer satellite imagery. J Med Entomol 35:989–995

Foody GM (2006) GIS: health applications. Prog Phys Geogr 30(5):691–695

Glass GE, Cheek JE, Patz JA, Shields TM, Doyle TJ, Thoroughman DA, Hunt DK, Enscore RE, Gage KL, Irland C, Peters CJ, Bryan R (2000) Using remotely sensed data to identify areas at risk for hantavirus pulmonary syndrome. Emerg Infect Dis 6(3):238

Glass GE, Schwartz BS, Morgan JM, Johnson DT, Noy PM, Israel E (1995) Environmental risk factors for Lyme disease identified with geographic information systems. Am J Public Health 85(7):944–948

Guerra M, Walker E, Jones C, Paskewitz S, Cortinas MR, Stancil A et al (2002) Predicting the risk of lyme disease: habitat suitability for Ixodes scapularis in the North Central United States. Emerg Infect Dis 8:289–295

Guo-Jing G, Chen H, Lin D, Hu G, Wu X, Li D et al (2002) A method of rapid identification snail habitat in marshland of Poyang Lake region by remote sensing. Chin J Parasit Dis 15:291–296

Hay SI, Snow RW, Rogers DJ (1998) Predicting mosquito habitat to malaria seasons using remotely sensed data: practice, problems and perspectives. Parasitol Today 14:306–313

Hay SI, Tatem AJ, Graham AJ, Goetz SJ, Rogers DJ (2006) Global environmental data for mapping infectious disease distribution. Adv Parasitol 62:37–77

Hendricks G, LaRocque S, Reid R, Wint W (2001) Spatial trypanosomiasis management: from data-layers to decision making. Trends Parasitol 17(1):35–41

Hendrickx G, Nepala A, Rogers D, Bastiaensen P, Slingenbergh J (1999) Can remotely sensed meteorological data significantly contribute to reduce costs of tsetse surveys? Mem Inst Oswaldo Cruz 94:273–276

Hennekens CH, Buring JE (1987) Epidemiology in medicine. Lippincott Williams & Wilkins

Hollingsworth TD, Pulliam JRC, Funk S, Truscott JE, Isham V, Lloyd AL (2015) Seven challenges for modelling indirect transmission: vector-borne diseases, macroparasites and neglected tropical diseases. Epidemics 10:16–20. https://doi.org/10.1016/j.epidem.2014.08.007

Honório NA, Codeço CT, Alves FC, Magalhães MAFM, Lourençode-Oliveira R (2009) Temporal distribution of Aedes aegypti in different districts of Rio de Janeiro, Brazil, measured by two types of traps. J Med Entomol 46:1001–1014

Hu W, Tong S, Mengersen K, Oldenburg B, Dale P (2005) Spatial and temporal patterns of Ross River virus in Brisbane, Australia. Arbovirus Res Aust 9:128–136

Huang Z, Das A, Qiu Y, Tatem AJ (2012) Web-based GIS: the vector-borne disease airline importation risk (VBD-AIR) tool. Int J Health Geogr 11:33

Jeganathan C, Khan SA, Chandra R et al (2001) Characterisation of malaria vector habitats using remote sensing and GIS. J Indian Soc Remote Sens 29:31. https://doi.org/10.1007/BF02989911

Kabatereine NB, Brooker S, Tukahebwa EM, Kazibwe F, Onapa AW (2004) Epidemiology and geography of Schistosoma mansoni in Uganda: implications for planning control. Tropical Med Int Health 9(3):372–380

Kalluri S, Gilruth P, Rogers D, Szczur M (2007) Surveillence of arthropod vector borne infectious diseases using remote sensing techniques: a review. PLoS Pathog 3(10):1361–1371

Kaya S, Pultz TJ, Mbogo CM, Beier JC, Mushinzimana E (2002) The use of radar remote sensing for identifying environmental factors associated with malaria risk in Coastal Kenya. In: International geoscience and remote sensing symposium (IGARSS'02), Toronto

Keegan L, Dushoff J (2014) Analytic calculation of finite-population reproductive numbers for direct- and vector-transmitted diseases with homogeneous mixing. Bull Math Biol 76(5):1143–1154. https://doi.org/10.1007/s11538-014-9950-x

Kesari S, Bhunia GS, Chatterjee N, Kumar V, Mandal R, Das P (2013) Appraisal of phlebotomus argentipes habitat suitability using a remotely sensed index in the Kala-azar endemic focus of Bihar, India. Memórias do Instituto Oswaldo Cruz 108(2):197–204

Kesari S, Bhunia GS, Kumar V, Jeyaram A, Ranjan A, Das P (2011) A comparative evaluation of endemic and non-endemic region of visceral leishmaniasis (Kala-azar) in India with ground survey and space technology. Memórias do Instituto Oswaldo Cruz 106(5):515–523

King RJ, Campbell-Lendrum DH, Davies CR (2004) Predicting geographic variation in Cutaneous Leishmaniasis, Colombia. Emerg Infect Dis 10:598–607

Kirk MD, Pires SM, Black RE, Caipo M, Crump JA, Devleesschauwer B, Döpfer D, Fazil A, Fischer-Walker CL, Hald T, Hall AJ (2015) World Health Organization estimates of the global and regional disease burden of foodborne bacterial, protozoal, and viral diseases, 2010: a data synthesis. PLoS Med 12(12):e1001921

Kiszewski A, Mellinger A, Spielman A, Malaney P, Sachs S (2004) A global index representing the stability of malaria Transmission. Am J Trop Med Hyg 70:486–498

Kitron U, Kazmierczak JJ (1997) Spatial analysis of the distribution of lyme disease in Wisconsin. Am J Epidemiol 145:558–566

Kitron U, Otieno LH, Hungerford LL, Odulaja A, Brigham WU, Okello OO et al (1996) Spatial analysis of the distribution of tsetse flies in the Lambwe Valley, Kenya, using Landsat Tm satellite imagery and GIS. J Anim Ecol 65:371–380

Lindsay SW, Thomas CJ (2000) Mapping and estimating the population at risk from lymphatic filariasis in Africa. Trans R Soc Trop Med Hyg 94:37–45

MacMahon B, Pugh TF (1970) Epidemiology; principles and methods. Little & Brown, Boston

Malaviya P, Picado A, Singh SP, Hasker E, Singh RP, Boelaert M, Sundar S (2011) Visceral leishmaniasis in Muzaffarpur district, Bihar, India from 1990 to 2008. PLoS One 6(3): e14751. https://doi.org/10.1371/journal.pone.0014751

Martin C, Curtis B, Fraser C, Sharp B (2002) The use of GIS-based malaria information system for malaria research and control in South Africa. Health and Place 8(4):227–236

Masuoka PM, Claborn DM, Andre RG, Nigro J, Gordon SW, Klein TA, Kim H (2003) Use of IKONOS and Landsat for malaria control in the Republic of Korea. US Army Res, Paper, p 336

Miranda C, Massa JL, Marques CA (1996) Occurrence of American Cutaneous Leishmaniasis by remote sensing satellite imagery in an urban area of Southeastern Brazil. Rev Saúde Pública 30:433–437

Miranda ML, Dolinoy DC (2005) Using GIS-based approaches to support research on neurotoxicants and other children's environmental health threats. Neurotoxicology 26(2): 223–228

Mohan VR, Naumova EN (2014) Temporal changes in land cover types and the incidence of malaria in Mangalore. India. Int J Biomed Res 5(8):494–498

Mollalo A, Alimohammadi A, Shahrisvand M, Shirzadi MR, Malek MR (2014) Spatial and statistical analyses of the relations between vegetation cover and incidence of cutaneous leishmaniasis in an endemic province, northeast of Iran. Asian Pac J Trop Dis 4:176–180

Muhar A, Dale PER, Thalib L, Arito E (2000) The spatial distribution of Ross River virus infections in Brisbane: significance of residential location and relationships with vegetation types. Environ Health Prev Med 4:184–189

Muller G, Grebaut P, Gouteux JP (2004) An agent-based model of sleeping sickness: simulation trials of a corest focus in southern Cameroon. CR Biolog 327:1–11

Nieto P, Malone JB, Bavia ME (2006) Ecological niche modeling for visceral leishmaniasis in the state of Bahia, Brazil, using genetic algorithm for rule-set prediction and growing degree day-water budget analysis. Geospatial Health 1(1):115–126

Odoi A, Martin SW, Michel P, Holt J, Middleton D, Wilson J (2003) Geographical and temporal distribution of human giardiasis in Ontario, Canada. Int J Health Geogr 2(1):5

Okorie PN, Marshall JM, Akpa OM, Ademowo OG (2014) Perceptions and recommendations by scientists for a potential release of genetically modified mosquitoes in Nigeria. Malar J 13:154

Omumbo JA, Hay SJ, Goetz SJ, Snow RW, Rogers DJ (2002) Updating historical maps of malaria transmission intensity in East Africa using remote sensing. Photogram Eng Remote Sens 68:161–166

Palaniyandi M (2014) The red and infrared IRS WiFS satellite data for mapping of malaria and JE vector mosquito breeding habitats. J Geophys Remote Sens 3:126. https://doi.org/10.4172/2169-0049.1000126

Palaniyandi M (2012) The role of remote sensing and GIS for spatial prediction of vector-borne diseases transmission: a systematic review. J Vector Borne Dis 49(4):197–204

Paul SK, Jeyaram A, Jayaraman V (2006) Application of remote sensing and GIS in identifying and mapping sandfly distribution in endemic and non-endemic Kala-azar foci in Bihar and Jharkhand. In: Proceedings of the 57th AIAA International Astronautical Congress (IAC'06), pp 2372–2387

Pellerin C (2006) United States updates global positioning system technology. http://www.america.gov/st/washfileenglish/2006/February/20060203125928lcnirellep0.5061609.html

Picheral HE (1994) Place, space, and health. Soc Sci Med 39:1589–1590

Polack S, Brooker S, Kuper H, Mariotti S, Mabey D, Foster A (2005) Mapping the global distribution of trachoma. Bull World Health Organ 83:913–919

Raban MZ, Dandona R, Dandona L (2009) Essential health information available for India in the public domain on the internet. BMC Publ Health 9:208

Raso G, Matthys B, N'Goran EK, Tanner M, Vounatsou P, Utzinger J (2005) Spatial risk prediction and mapping of Schistosoma mansoni infections among schoolchildren living in western Côte d'Ivoire. Parasitology 131:97–108

Ratmanov P, Mediannikov O, Raoult D (2013) Vector borne diseases in West Africa: geographic distribution and geospatial characteristics. Trans Roy Soc Trop Med Hyg 1–12

Rejmankova E, Pope KO, Roberts DR, Lege MG, Andre R, Greico J et al (1998) Characterization and detection of Anopheles vestitipennis and Anopheles punctimacula (Diptera: Culicidae) larval habitats in Belize with field survey and SPOT satellite imagery. J Vector Ecol 23:74–99

Rincón-Romero ME, Londoño JE (2009) Mapping malaria risk using environmental and anthropic variables. Rev Bras Epidemiol 12(3):338–354

Robinson RA, Lawson B, Toms MP, Peck KM, Kirkwood JK, Chantrey J, Clatworthy IR, Evans AD, Hughes LA,Hutchinson OC, John SK, Pennycott TW, Perkins MW, Rowley PS, Simpson VR, Tyler KM, Cunningham AA (2010) Emerging infectious disease leads to rapid population declines of common british birds. PLoS One 5(8):e12215. https://doi.org/10.1371/journal-.pone.0012215

Robinson TP, Rogers D, Williams B (1997) Mapping tsetse habitat suitability in the common fly belt of Southern Africa using multivariate analysis climate and remotely sensed vegetation data. Med Vet Entomol 11:235–245

Rodriguez AD, Rodriguez MH, Hernandez JE, Dister SW, Beck LR, Rejmankova E et al (1996) Landscape surrounding human settlements and malaria mosquito abundance in Southern Chiapas, Mexico. J Med Entomol 33:39–48
Rogers DJ (2006) Models for vectors and vector-borne diseases. Adv Parasitol 62:1–35
Rogers DJ (2000) Satellites, space, time and the African trypanosomiases. Adv Parasitol 47:129–171
Rogers DJ (1991) Satellite imagery tsetse and trypanosomiasis in Africa. Prev Vet Med 11:201–220
Rogers DJ, Hay SI, Packer MJ (1996) Predicting the distribution of tsetse flies in West Africa using temporal fourier processed meteorological satellite data. Ann Trop Med Parasitol 90:225–241
Rogers DJ, Randolph SE, Snow RW, Hay SI (2002) Satellite imagery in the study and forecast of malaria. Nature 415:710–715
Rutto JJ, Karuga JW (2009) Temporal and spatial epidemiology of sleeping sickness and use of geographical information system (GIS) in Kenya. J Vector Borne Dis 46(1):18–25
Saathoff E, Olsen A, Kvalsvig JD, Appleton CC, Sharp BL, Kleinschmidt I (2005) Ecological covariates of ascaris lubricoides infection in schoolchildren from rural KwaZulu-Natal, South Africa. Trop Med Int Health 10(5):412–422
Salomon OD, Orellano PW, Quintana MG, Pérez S, Sosa Estani S, Acardi S, Lamfri M (2006) Transmisión de la leishmaniasis tegumentaria en Argentina. Med (B Aires) 66:211–219
Saraiva L, Andrade-Filho JD, Falcão AL, Carvalho DAA, Souza CM, Freitas CM, Lopes CRG, Moreno EC, Melo MN (2011) Phlebotominae fauna (Diptera: Psychodidae) in an urban district of Belo Horizonte, Brazil, endemic for visceral leishmaniasis: Characterization of favored locations as determined by spatial analysis. Acta Tropica 117:137–145
Shah NH, Gupta J (2013) SEIR model and simulation for vector borne diseases. Appl Math 4:13–17
Sherchand JB, Obsomer V, Thakur GD, Hommel M (2003) Mapping of lymphatic filariasis in Nepal. Filaria J 2(1):7
Shimabukuro PHF, da Silva TRR, Fonseca FOR, Baton LA, Galati EAB (2010) Geographical distribution of American cutaneous leishmaniasis and its phlebotomine vectors (Diptera: Psychodidae) in the state of São Paulo, Brazil. Parasites Vectors 3:121
Shirayama Y, Phompida S, Shibuya K (2009) Geographic information system (GIS) maps and malaria control monitoring: intervention coverage and health outcome in distal villages of Khammouane province. Laos Malar J 8:217
Simarro PP, Cecchi G, Paone M, Franco JR, Diarra A, Ruiz JA, Fèvre EM, Courtin F, Mattioli RC, Jannin JG (2010) The Atlas of human African trypanosomiasis: a contribution to global mapping of neglected tropical diseases. Int J Health Geogr 9:57
Sipe NG, Dale P (2003) Challenges in using geographic information systems (GIS) to understand and control malaria in Indonesia. Malar J 2:36
Sithiprasasna R, Lee WJ, Ugsang DM, Linthicum KJ (2005) Identification and characterization of larval and adult anopheline mosquito habitats in the Republic of Korea: potential use of remotely sensed data to estimate mosquito distributions. Int J of Health Geogr 4:17
Strom SR (2002) Charting a course toward global navigation. Retrieved 12 Jul 2008. http://www.aero.org/publications/crosslink/summer2002/01.html
Sudhakar S, Srinivas T, Palit A, Kar SK, Battacharya SK (2006) Mapping of risk prone areas of Kala-azar (Visceral leishmaniasis) in parts of Bihar state, India: an RS and GIS approach. J Vect Borne Dis 43:115–122
Symeonakis E, Robinson T, Drake N (2007) GIS and multiple-criteria evaluation for the optimization of tsetse fly eradication programmes. Environ Monit Assess 124:89–103
Thompson DF, Malone JB, Harb M, Faris R, Huh OK, Buck AA et al (1996a) Bancroftian filariasis distribution and diurnal temperature differences in the Southern Nile Delta. Emerg Infect Dis 2:234–235
Thompson MC, Connor SJ, Milligan PJ, Flasse SP (1996b) The ecology of malaria as seen from earth-observation satellites. Ann Trop Med Parasitol 90:243–264
Thompson MC, Elnaiem DA, Ashford RW, Connor SJ (1999) Towards a kala azar risk map for Sudan: map ping the potential distribution of Phlebotomus orientalis using digital data of environmental variables. Trop Med Int Health 4:105–113

Tobler WR (1969) Geographical filters and their inverses. Geogr Anal 1(3):234–253

Townshend JRG, Justice CO (1986) Analysis of the dynamics of African vegetation using the normalized difference vegetation index. Int J Remote Sens 7:1435–1445

Tran A, Ponçon N, Toty C, Linard C, Guis H, Ferré JB, Lo Seen D, Roger F, de la Rocque S, Fontenille D, Baldet T (2008) Using remote sensing to map larval and adult populations of Anopheles hyrcanus (Diptera: Culicidae) a potential malaria vector in Southern France. Int J Health Geogr 7(1):9

Troyo A, Fuller DO, Calderón-Arguedas O, Beier JC (2008) A geographical sampling method for surveys of mosquito larvae in an urban area using high-resolution satellite imagery. J Vector Ecol 33(1):1–7

Tuthill K (2003) John snow and the broad street pump: on the trail of an epidemic. Cricket 31(3): 23–31

Upadhyayula SM, Mutheneni SR, Kadiri MR, Kumaraswamy S, Nagalla B (2012) A cohort study of lymphatic filariasis on socio economic conditions in Andhra Pradesh, India. PLoS One 7(3): e33779. https://doi.org/10.1371/journal.pone.0033779

Vally H, Peel M, Dowse GK, Cameron S, Codde JP, Hanigan I, Lindsay MDA (2012) Geographic Information Systems used to describe the link between the risk of Ross River virus infection and proximity to the Leschenault estuary, WA. Aust N Z J Public 36(3):229–235

Werneck GL, Maguire JH (2002) Spatial modeling using mixed models: an ecologic study of visceral leishmaniasis in Teresina, Piauí State, Brazil. Cad Saúde Pública 18:633–637

Wijegunawardana NDAD, Silva Gunawardene YIN, Manamperi A, Senarathne H, Abeyewickreme W (2012) Geographic information system (GIS) mapping of lymphatic filariasis endemic areas of Gampaha district, Sri lanka based on epidemiological and entomological screening, Southeast Asian. J Trop Med Public Health 43(3):557–566

World Health Organization (WHO) (1999) Geographical information systems (GIS). Wkly Epidemiol Rec 74:281–285

Zeilhofer P, Santos E, Ribeiro A, Miyazaki R, Santos M (2007) Habitat suitability mapping of Anopheles darlingi in the surroundings of the Manso hydropower plant reservoir, Mato Grosso, Central Brazil. Int J Health Geogr 6(1):7

Chapter 2
Spatial Database for Public Health and Cartographic Visualization

2.1 Foundation of Spatial Data

The database is the foundation of GIS (Worboys and Duckham 2004). GIS application entails access to geo-spatial data (e.g., attained by observation and aspect of an object or events referenced to their position in that space). Spatial data framework is very much significant to the functioning, planning, maintenance, and management of data, in which other data layers are tied to geographical frame of reference. National Academy of Science (1995) stated that foundations of spatial data are "*the minimal directly observable or recordable data to which other data are spatially referenced*". The Australia New Zealand Land Information Council (ANZLIC) defines foundation spatial data as: '*the authoritative geographic information that underpins, or can add significant value to, any other information. It supports evidence-based decisions across government, industry and the community*'. However, certain data may indeed be appropriate with more than one theme; for example, roads transport infrastructure and administrative boundaries.

In creating spatial database for public healthcare service in GIS platform, the linkage among the data layers is knotted together by their shared geographical location. It is somewhat difficult to connect data layers in straight way to several types of data that can serve as foundation data, for example, orthoimagery (digital dataset derived from the corrected aerial photograph), elevation (digital files defining the height of the land surface and depth below the water surface), transportation (road, railways), hydrography (rivers, lakes, reservoir), and cadastral information (boundaries defining land ownership). Foundation spatial data must have the following properties:

- Necessary for public welfare and safety
- Acute for a general function

G. S. Bhunia and P. K. Shit, *Geospatial Analysis of Public Health*,
https://doi.org/10.1007/978-3-030-01680-7_2

- Contributes considerably to social, economic and environmental sustainability
- Innovative information added by government, industry, research and academic sectors.

There are several assistances for the national framework of spatial data in public health:

- Lessening the costs allied with data copying across agencies, and the storage and access of data
- Speedy and easy way to access basic data that can be simply assumed and employed by various operators to which extrageographically referenced data can be accumulated
- Advances user assurance and fulfillment by creating available consistent and regular foundation spatial data
- Enhanced policy making by local, state and government agencies, private sector
- Higher levels of invention and rivalry ensuing in better products and amenities being accessible in market
- Improved interpretability through the practice of generally reliable datasets for cross managerial analysis, processes and policy making.

2.2 Scale of Public Health Data

2.2.1 Worldwide Level

Public health investigation is the orderly gathering, analysis, understanding, and propagation of health data on a continuing basis, to increase acquaintance of the pattern of disease incidence in order to switch and preclude disease in the community. Various spatial analyses of public health data employ health outcome data composed and abridged by government agencies, state health departments, and then released for free use. The public health data (e.g., birth, death, clinical outcomes, and location of health centers) may aid as an early warning system for imminent public health difficulties; and observe and elucidate the epidemiology of health problems, and to notify public health strategy and policies. The data types comprise a variability of collection procedures and purposes, and consequently, differ in terms of accessibility, spatial coverage, and spatial support. Here, we abridge numerous types of health data poised by several national and local governments, with special consideration to spatial aspects of different data sources. A lot of data sources recorded below are freely accessible through interaction with the organization or administrations that are accountable for the data. Global Foodborne Infections Network (GFN) (http://www.who.int/gho/en/), WHO Global InfoBase: health statistics on chronic disease (https://apps.who.int/infobase/), HIV surveillance, estimations and monitoring and evaluation (http://www.who.int/hiv/topics/me/en/) epitomizes a preliminary effort to carry together obtainable data on different aspect of health and health-related indicators.

2.2.2 Country Level Public Health Data: An Example

Many countries in the world provide free secondary data for accessing and analysing the data for long-term research and policy making. In 1998, CEHAT released over 500 variables of public health data with its own software programme in the DOS environment from 1951 to the latest available years (http://www.cehat.org/publications/rm06.html). Some of the state-wise time series evidence provided in the database includes: general profile, health indicator, health infrastructure, health human power, health finances etc.

National Informatics Centre (NIC), Department of Electronics & Information Technology, Ministry of Communications & Information Technology, and Government of India develop a database to provide the state-wise statistics of mental hospitals like number of hospitals, beds and patients (https://data.gov.in/-keywords/public-health-statistics). Moreover, Institute for Health Metrics and Evaluation (IHME), University of Washington develop Global Burden of Disease (GBD) tool to enumerate health forfeiture from hundreds of infections, damages, and risk factors, so that health schemes can be enhanced and inequalities can be eradicated (http://www.who.int/healthinfo/global_burden-_disease/gbd/en/). World Health Organization (WHO) comprises data on about 600 health indicators for Member States of the WHO European Region (http://data.euro.who.int/hfadb/), with basic demographic and socio-economic parameters; standard of living- and environment-interconnected indicators; morbidity, mortality and disability; hospital liberations; and health care assets, deployment and expenditure. Kaiser Family Foundation (KFF), US provided Global data on HIV/AIDS, malaria, TB and other key health and socio-economic indicators (http://www.globalhealthfacts.com/).

Centers for Disease Control and Prevention, USA provides GIS data, like the spatial aspects of health and illness (https://www.cdc.gov/gis/gis-at-cdc.htm). The Vermont Department of Health practices GIS technology to sustain a wide range of public health activities (e.g., health surveillance, environmental health, public health preparedness, health promotion and disease prevention, maternal child health, local health/rural health etc.) that can be observed by mobile devices, such as smart phones and tablet computers, by creating maps and reports that are tailored to mobile devices (http://www.health-vermont.gov/GIS/). In India, National Resource Data Management System (NRDMS), Department of Science and Technology, Govt. of India develop geospatial technological tools to concern numerous facets associated to public health care.

2.2.3 Regional Level Public Health Data: An Example

The National Vector Borne Disease Control programme (NVBDCP) is providing information of vector borne diseases i.e. Malaria, Dengue, Lymphatic Filariasis, Kala-azar, Japanese Encephalitis and Chikungunya in India at regional level (http://www.nvbdcp.gov.in/). The Christian Medical College, Vellore (CMC) is working with the Environmental Systems Research Institute (ESRI), California to offer supplementary insights for better understanding the intricacies in disease transmission and disparities in health among communities in Vellore district (http://www.cmch-vellore.edu/static/directorate/-awards/pdf/gis.pdf). The country level availability of public health database is illustrated in Table 2.1.

2.3 Digital Cartographic Data

The accessibility of geographically referenced data is increasing to increase at rapid pace. Substantial efforts have been completed to coordinate and normalize the production and dissemination of digital cartographic data and most of these data are now readily available open on the internet (Table 2.2). Internet is medium for finding the existing digital cartographic data from non-profit organizations and private companies. The US Geological survey (USGS) is a major provider of GIS data in the United States (https://www.usgs.gov/) and the entire world. Its website (http://glovis.usgs.gov/) provides corridors to USGS national mapping and satellite data and various thematic layers. However, data on the internet may be in a format that is irreconcilable with GIS software used for a project, or to be functioning for a project. Global Map (https://www.iscgm.org/gmd/) developed by the International Steering Committee on Global Mapping is consistent with GIS layers covering the whole globe at 1 km resolution including: transportation, elevation, drainage, vegetation, administrative boundaries, land cover, land use and population centered at national and regional level.

DIVA-GIS (http://www.diva-gis.org/about) developed by Robert Hijmans, is a free computer program for mapping and analysis of geographic data, delivers the GIS layers (includes administrative areas, inland water, roads and railways, elevation, land cover, population and climate) publicly for the entire world in both vector and GRID format. United Nations Environment Programme Global Environment Outlook (UNEP-GEO) data provides wide range of data including global forest cover, global potential evapotranspiration, global average monthly temperatures, dams, watershed boundaries and much more (http://geodata.grid.unep.ch/). Its online database clasps more than 500 different variables, as national, sub-regional, regional and global statistics.

GeoNetwork (http://www.fao.org/geonetwork/srv/en/main.home) offers both the human and physical database at national and regional level. The GIS Data Depot (http://data.geocomm.com/) is a new geospatial data portal established by an

Table 2.1 Spatial resolution of public health data

Name	Characteristics of data	Relevant paper/ sponsored agency	Website
NIH Data Sets	A large collection of data sets on various aspects of health care hosted by the NIH	U.S. National Library of Medicine	http://www.nlm.nih.gov/hsrinfo/datasites.html
The Human Mortality Database	Provide detailed mortality and population data over 38 countries	Department of Demography at the University of California, Berkeley, USA, and Max Planck Institute for Demographic Research in Rostock, Germany	http://www.mortality.org/
World Bank Health Care Data Sets	Improving health is central to the Millennium Development Goals, and the public sector is the main provider of health care in developing countries. To reduce inequities, many countries have emphasized primary health care, including immunization, sanitation, access to safe drinking water, and safe motherhood initiatives	The World Bank, IBRD, IDA	http://data.worldbank.org/country
World Health Organization Data Sets	The World Health Statistics series is WHO's annual compilation of health statistics for its 194 Member States. World Health Statistics 2016 focuses on the proposed health and health-related Sustainable Development Goals (SDGs) and associated targets	World Health Organization (WHO)	http://apps.who.int/ghodata/
Behavioral Risk Factor Surveillance System (BRFSS)	Health care survey data from 1984 to 2009-contains both summary statistics and raw data	http://www.cdc.gov/brfss/pubs/index.htm	http://www.cdc.gov/brfss/

(continued)

Table 2.1 (continued)

Name	Characteristics of data	Relevant paper/ sponsored agency	Website
Open Government Data (OGD) Platform India	To Provide Household House list information under Annual Health Survey (AHS) data, linical, Anthropometric & Bio-chemical (CAB) Survey data, and key indicator of HMIS at district level	Govt. of India	https://data.gov.in/
National Information Center on Health Services Research and Health Care Technology (NICHSR)	The collection, storage, analysis, retrieval, and dissemination of information on health services research, clinical practice guidelines, and on health care technology, including the assessment of such technology	U.S. National Library of Medicine	https://www.nlm.nih.gov/hsrinfo/datasites.html
National Health Portal	To provide healthcare related information to the citizens of India and to serve as a single point of access for consolidated health information	NHP Admin, Govt. of India	http://www.nhp.gov.in/about-national-health-portal_pg

e-governance initiative and provides data and information commercially in various vector data format (DRG, DEM, NWI, DLG, and DOQ). However, one of the main objectives of these entire open source databases is to swell collaborative partnership at all levels of government to aid control investments in geospatial data and to lessen repetition of data. Many GIS companies (ESRI, GDT, Tele Atlas) are betrothed in software development, technical service, consulting and data production.

2.4 Database Integration

Data integration means combining data derived from various sources and delivering users with a united view of these data (Fig. 2.1). GIS specialist is used to integrate the data from various sources based on location, and have a clear concept of the data layers. Several common problems get up during networking and integrating the spatial datasets.

Table 2.2 Global open sources digital cartographic data and spatial databases for purposes of geospatial analysis and cartographic mapping

Name	Characteristics of data	Data type	Scale	Providing agency	Website
Natural Earth	Public domain map dataset	Vector and raster	1:10 million, 1:50 million, 1:110 million	North American Cartographic Information Society	https://en.wikipedia.org/wiki/Natural_Earth
Global Map	All the Earth's land cover (transportation, elevation, drainage, vegetation, administrative boundaries, land cover, population centres, and land use)	Thematic layers	1 km resolution	Ministry of Construction, Japan (MOC)	http://www.gsi.go.jp/kankyochiri/globalmap_e.html
UNEP Environmental Data Explorer	Global forest cover, global potential evapotranspiration, global average monthly temperatures, dams, watershed boundaries	Maps, graphs, data tables	National, sub-regional, regional	United Nations Environment Programme	http://geodata.grid.unep.ch/
Global Self-consistent, Hierarchical, High-resolution Geography Database (GSHHG)	World Vector Shorelines (WVS) CIA World Data Bank II (WDBII) Atlas of the Cryosphere (AC)	Vector		Dr. Paul Wessel (University of Hawai'i), Dr. Walter H. F. Smith (NOAA Laboratory for Satellite Altimetry)	http://www.soest.hawaii.edu/pwessel/gshhg/
ASTER GDEM	Global elevation data derived from Advanced Spaceborne Thermal Emission and Reflection Radiometer satellite images	Raster	30 m	Ministry of Economy, Trade, and Industry (METI) of Japan and the United States National Aeronautics and Space Administration (NASA)	http://asterweb.jpl.nasa.gov/gdem.asp

(continued)

Table 2.2 (continued)

Name	Characteristics of data	Data type	Scale	Providing agency	Website
Open Topography	High-resolution, Earth science-based, topography data, and related tools and resources.	Raster, point cloud data	30–90 m DEM	San Diego Supercomputer Center	http://opentopo.sdsc.edu/lidar
NCAR GIS Climate Change Scenarios	Includes data used by the IPCC	Vector and raster		National Center for Atmospheric Research	http://gisclimatechange.ucar.edu/
IRI/LDEO Climate Data Library	300 datasets from various climate models	Vector and raster		The Earth Institute and Lamont–Doherty Earth Observatory	http://iridl.ldeo.columbia.edu/
Global Climate Monitor	Climate web viewer containing accessible climatic information from 1901	Grided data		Global Precipitation Climatology Centre	http://www.globalclimatemonitor.org/
HydroSHEDS	Data and maps based on Shuttle Elevation Derivatives includes river networks, watershed boundaries, drainage directions, and flow accumulations	Vector and raster	Multiple scales	Conservation Science Programme of World Wildlife Fund (WWF)	http://hydrosheds.cr.usgs.gov/index.php
USGS Land Cover Institute	Global land cover datasets; USGS releases 1973–2000 time-series of land-use/land-cover for the conterminous U.S.	Raster	Multiple scales	U.S. Department of the Interior, U.S. Geological Survey	http://landcover.usgs.gov

(continued)

Table 2.2 (continued)

Name	Characteristics of data	Data type	Scale	Providing agency	Website
Global 200	Earth's terrestrial, freshwater, and marine ecoregions that harbor exceptional biodiversity and are representative of its ecosystems. This process yielded 238 ecoregions–the Global 200–comprised of 142 terrestrial, 53 freshwater, and 43 marine priority ecoregions	Vector		WWF	http://www.worldwildlife.org/publications/global-200
GLWD	Lakes and wetlands on a global scale on (1) large lakes and reservoirs, (2) smaller water bodies, and (3) wetlands	Vector	1:1–1:3 million resolution	WWF and Center for Environmental Systems Research, University of Kassel, Germany	http://www.worldwildlife.org/pages/global-lakes-and-wetlands-database
Conservation GIS Data	TNC Lands and Waters (properties/preserves, easements and leases), Ecoregional Portfolio, Terrestrial Ecoregions of the World, Freshwater Ecoregions of the World, and Marine Ecoregions of the World	KML formats, advanced vector (REST)		The Nature Conservancy	http://maps.tnc.org/gis_data.html
Pilot Analysis of Global Ecosystems: Forest Ecosystems	Percentage tree-cover, population density and tree cover, share of wood in fuel consumption, etc.	Raster, map		World Resources Institute	http://www.wri.org/resources

(continued)

Table 2.2 (continued)

Name	Characteristics of data	Data type	Scale	Providing agency	Website
SoilGrids1 km—soil property and class maps	Soil pH, texture (sa, si, cl), organic carbon and more for 6 depth layers up to 2 m depth	Raster	1 km resolution	International Soil Science Society (ISSS) and a resolution of the United Nations Educational, Scientific and Cultural Organization (UNESCO)	http://soilgrids1km.isric.org/
Harmonized World Soil Database	Combines regional and national soil databases (pH, depth, and texture parameters) and maps	Vector	30 arc-second resolution	Food and Agriculture Organization of the United Nations	http://webarchive.iiasa.ac.at/Research/LUC/External-World-soil-database/HTML/HWSD_Data.html?sb=4
Datasets from the United Nations Environment Programme	Includes global ecoregions, wetlands, distribution of coral reefs, mangroves, and more				http://datadownload.unep-wcmc.org/datasets
Mineral Resource Data System	Metallic and nonmetallic mineral resources throughout the world. Includes names, location, commodity, description, geologic characteristics, production, reserves, and more	Raster and vector		USGS	http://mrdata.usgs.gov/
Global Administrative Areas	Administrative areas in this database are countries and lower level subdivisions such as provinces,	shapefile, ESRI geodatabase, RData, and		Robert Hijmans, University of California, Julian Kapoor and John Wieczorek (Berkeley	http://gadm.org/

(continued)

Table 2.2 (continued)

Name	Characteristics of data	Data type	Scale	Providing agency	Website
	departments, bibhag, bundeslander, daerahistimewa, fivondronana, krong, landsvæðun, opština, sous-préfectures, counties, and thana	Google Earth kmz format		Museum of Vertebrate Zoology), the International Rice Research Institute (Nel Garcia, Aileen Maunahan, ArnelRala) and the University of California, Davis (Alex Mandel), and with contributions of many others	
World Bank Geodata	Schooling and financial data, etc.	KML format			https://sourceforge.net/projects/googleworldbank/
Global Agriculture Lands	Agricultural inventory data (MODIS and SPOT image)	Raster (GeoTiff and GRID formats)	1 km resolution	NASA	http://sedac.ciesin.columbia.edu/data/collection/aglands/sets/browse
Historic Croplands Dataset, 1700–1992	Historical changes in global land cover: croplands from 1700 to 1992	NetCDF and ArcINFO ASCII snapshots	0.5° resolution	Ramankutty and Foley (1999)	https://nelson.wisc.edu/sage/data-and-models/historic-croplands/index.php
Global Reservoir and Dam (GRanD) Database	Reservoirs with a storage capacity > 0.1 cubic km	Vector		Lehner et al. (2011); Global Water System Project (GWSP) and by the Columbia University Center for International Earth Science Information Network (CIESIN)	http://sedac.ciesin.columbia.edu/data/set/grand-v1-dams-rev01/data-download

(continued)

Table 2.2 (continued)

Name	Characteristics of data	Data type	Scale	Providing agency	Website
Gridded Population of the World (GPW)	Population data (raw population, population density, historic, current and predicted)	Raster	30 arc-seconds (approximately 1 km at the equator)	United Nations Food and Agriculture Programme (FAO)	http://beta.sedac.ciesin.columbia.edu/data/collection/gpw-v4.
Global Roads Open Access Data Set (gROADS)	Roads between settlements are included, not residential streets	Vector (Geodatabase and Shp.)	1:250,000	Center for International Earth Science Information Network—CIESIN—Columbia University, and Information Technology Outreach Services—ITOS—University of Georgia. 2013	http://sedac.ciesin.columbia.edu/data/set/groads-global-roads-open-access-v1/data-download
OpenStreetMap	Crowdsourced data for the whole world (points of interest, buildings, roads and road names, ferry routes etc.)			Steve Coast	https://www.openstreetmap.org/#map=5/51.500/-0.100
World Port Index	Location and physical characteristics of, and the facilities and services offered by major ports and terminals world-wide	Vector		United States National Geospatial-Intelligence Agency	http://msi.nga.mil/NGAPortal/MSI.portal?_nfpb=true&_pageLabel=msi_portal_page_62&pubCode=0015
Viewfinderpanoramas Digital Elevation Model (DEM) repository	Global digital elevation model data	HGT format	3 arc-second	Jonathan de Ferranti	http://www.viewfinderpanoramas.org/dem3.html

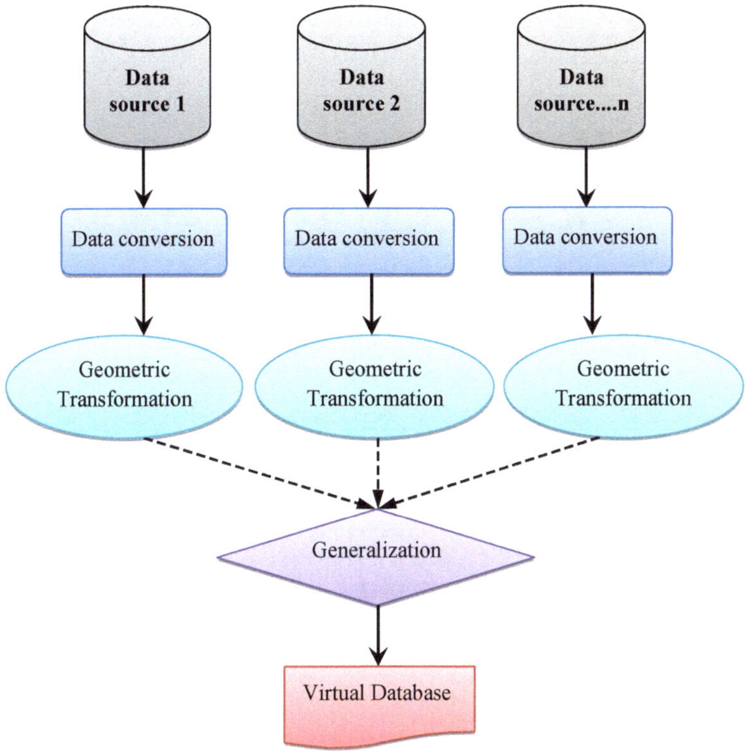

Fig. 2.1 Graphic representation of database integration

For example, data conversions provide geographical errors in creating and connecting spatial data sets. Data conversion in GIS includes importing/exporting, clipping and raster/vector format. For the procedure to work correctly, import/export of the data is essential. Import means that the data is converted from its original format into another format (e.g. IMG, TIFF, or GRID Stack etc.), that can be accessed directly for different application. Conversely, exporting (e.g., CSV, PDF, MS Excel etc.) indicates that the transfer application reformats the data for the getting application. In GIS, one can extract their study of interest of a mapped data set by cookie-cut features and attributes from the surrounding areas.

Another common problem arises in data integration is that the data on dissimilar projection system cannot be evocatively assimilated in GIS without supplementary processing to convey the data layer into analogous projection (Fig. 2.2). Collective in entirely GIS are measures for transforming co-ordinates so that they are reliable with those of the foundation data layer. The co-ordinate transformation includes translation (e.g., shifts its origin to a new location), rotation (e.g., rotates objects x and y axis from the origin), map projection (e.g., systematic arrangement of parallels and meridians on a plane surface) and rubber sheeting (e.g., surface is

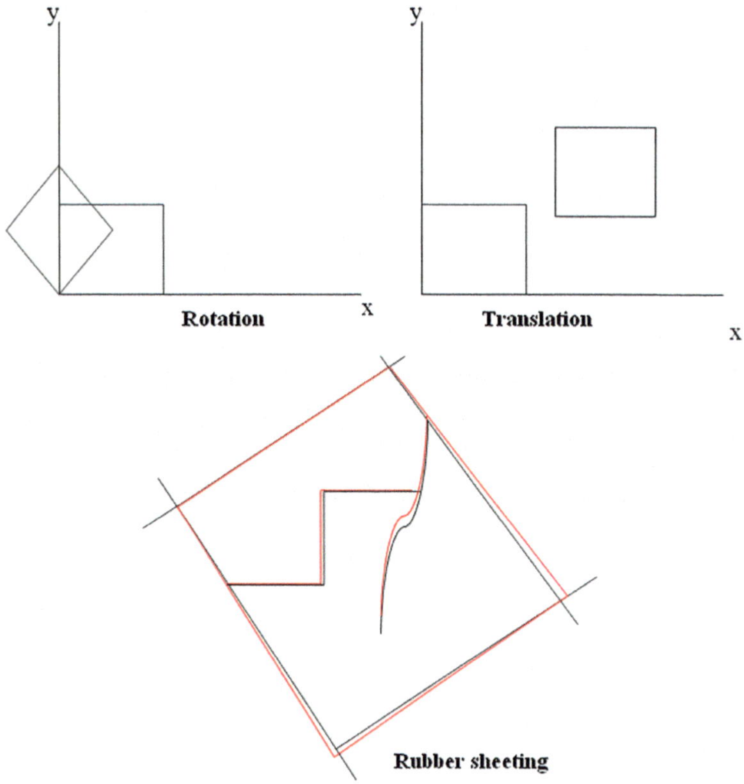

Fig. 2.2 Co-ordinate transformation techniques of GIS data

accurately overextended, affecting features using a piecewise transformation that conserves straight lines).

As such, data having dissimilar scale cannot be eloquently unified in a GIS without simplifying to the smallest scale. Most GIS packages comprise a sequence of cartographic and geographic technique for relating and integrating spatial dataset.

Many health databases are available only for geographical area like countries, ZIP codes, census zones. They comprise area, name, and a set of variables that pronounce the health events, populations, or additional attribute of the area. A spatial join is a GIS maneuver that attaches data from one feature layer's attribute table to another layer based on the spatial relationship (Table 2.3). The target features and the combined attributes from the join features are written to the output feature class. The spatial association is dogged by the geometry types of the input layers as well as the selected match option.

Table 2.3 Spatial relationships and the Spatial Join tool in ArcGIS software

Target feature	Join feature	INTERSECT	CONTAINS	WITHIN	CLOSEST
Point	Point	A join point is matched to a target point at the same location	Same as INTERSECT	Same as INTERSECT	A join point is matched to the nearest target point within the search radius
Point	Line	A join line is matched to a target point that it intersects	Not applicable	A join line is matched to a target point contained within the line	A join line is matched to the nearest target point within the search radius
Point	Polygon	A join polygon is matched to a target point that is inside or on the boundary of the polygon	Not applicable	A join polygon is matched to a target point that is inside the boundary of the polygon	A join polygon is matched to the nearest target point within the search radius
Line	Point	A join point is matched to a target line that it intersects	A join point is matched to a target line that contains it	Not applicable	A join point is matched to the nearest target line within the search radius
Line	Line	A join line is matched to a target line that it intersects. This includes lines that cross or are coincident	A join line is matched to a target line that completely contains it. Here the join line is coincident with the target line	A join line is matched to a target line that falls completely within the join line. Here the target line is coincident with the join line	A join line is matched to the nearest target line within the search radius
Line	Polygon	A join polygon is matched to a target line that it intersects	Not applicable	A join polygon is matched to a target line that is inside the boundary of the join polygon	A join polygon is matched to the nearest target line within the search radius
Polygon	Point	A join point is matched to a target polygon that it intersects or is inside of	A join point is matched to a target polygon that contains it	Not applicable	A join point is matched to the nearest target polygon

(continued)

Table 2.3 (continued)

Target feature	Join feature	INTERSECT	CONTAINS	WITHIN	CLOSEST
					within the search radius
Polygon	Line	A join line is matched to a target polygon that it intersects or is inside of	A join line is matched to a target polygon that contains it	Not applicable	A join line is matched to the nearest target polygon within the search radius
Polygon	Polygon	A join polygon is matched to a target polygon that it intersects or is inside of	A join polygon is matched to a target polygon that contains it	A join polygon is matched to a target polygon that is inside the boundary of the join polygon	A join polygon is matched to the nearest target polygon within the search radius

Data source ArcGIS v10.1, ESRI, Atlanta

Joining entails that each geographic area has a sole identifier, either a text or integer. Usually joining associates area-based health information to the conforming geographic areas in a foundation spatial database. However, spatial information is misplaced when address based health data are combined to areas than geocoded as points.

2.5 Public Health Data Sharing

Spatial data are the key components in GIS and the most important aspects of GIS analysis are database construction. Due to the advancement of Information Technology, these situations have changed as digital data clearing houses have become common place on the internet.

2.5.1 Localization of Spatial Data

Public health is one of the most important human activities producing very important but very complicated and often unintelligible spatial data. Health data and its accurate interpretation can have preventive effect or can map and control different epidemics. Current economic crisis places emphasis on cost saving of majority of states of the world. At the local level, data are used to observe population health. At national level to target intervention data are stored for resource distribution. And at the global level, health database is generated for the assessment of international liability of disease, the measurement of the growth in health and the development of universal funding agency. At this level principle for data sharing in global health is also set.

2.5.2 *Framework Data*

Most GIS data on the internet are the data that many organization regularly use for GIS activities, these are called as framework of data. As such, new GIS data can be created from variety of data sources. They include satellite images, field data, street address, text files with x and y co-ordinates and paper maps. Organizations may lay data in a standard format like Geography Markup Language (GML), the Extensible Markup Language (XML), syntax defined by the open geospatial consortium to express geographic features (Open Geospatial Consortium 2011) and deal out it on the internet. Conversely, the generating of spatial dataset is expensive, it is estimated that well over half the cost of GIS projects goes to database creation, updating and editing.

A second type of data sharing application involves, various agencies can publish and periodically update static map and they can maintain web-GIS (National Spatial Data Infrastructure (NSDI), Federal Geographic Data Committee). Several initiatives completely applied elucidations for sharing of health data such as international household survey network, the demographic and health survey (DHS), the multi indicator cluster survey (MICS) and international network for the continuous demographic evaluation (INDEPTH). As such, international agencies World Intellectual Property Organization (WIPO), World Health Organization (WHO), World Trade Organization (WTO) should lead to creation a political framework in the form of resolution or a treaty or operational guidelines for data sharing in Public health (Sankoh et al. 2013).

2.5.3 *Data from Private Companies*

Serving GIS analysis tools is another most important application of data sharing. It empowers users to edit online spatial database and accomplish analysis without setting GIS softwares. The South Carolina Department of Health and Environmental Controls ropes a geocoding facility accessible via a standard lone web interface (Shoultz 2005). These organizations' web service geocodes almost 100% of all data succumbed and outputs data in several co-ordinate systems.

2.5.4 *Limitations of Data Sharing*

There are some barriers of data sharing, like technical (public health data collection, language barrier, restrictive data format, technical solutions not available, lack of metadata and standards), motivational (no incentives, opportunity cost, possible criticism, disagreement on data use), economic (possible economic damages, lack of resources), political (lack of trust, restrictive policies, lack of guidelines), legal (Ownership copyright and protection of privacy), and ethical barriers (lack of proportionality and reciprocity (Van Panhuis et al. 2014). These fences for the most part are well tacit as part of strong contests in health information system capacity and endure to practice a major hindrance to the obtainability and usage of public health data.

2.5.5 *Benefits of Data Sharing*

Advantages in data sharing have been extensively documented—transparency, and collaboration, reproducibility and enquiries, cost-efficiency and stopping, severances, hastening of detection and intervention and savings lives through more competent and operational public health programs (Wartenberg and Thompson 2010a, b).

2.6 Data Visualization and Exploration

Visualization of data was taking place in the 2nd century C.E. with the preparation of data into row and columns and embryonic to the early quantitative illustrations in the 17th Century. The earliest instances of geographic visualization even dated back to the Stone Age with map-like wall paintings portraying the environments of our lineages. The competence of visual tools (imagery, maps and graphical products) and of the human visual treating scheme cannot be undervalued in terms of their significance for accepting the intricacies of spatial information. Photography endorsed for more precise recording and documentary apparatuses, particularly for the demonstration of landscape and terrain. The evolution of interest in visualizing the statistical landscape initiated to yield hold in the scientific community in the late sixteenth century, with researchers' concentration modifying from the physical to abstract world (Nixon 2011). Cartographic enquiry over last two decades has been anxious with systematizing traditional cartographic depictions and practices (Groop and Smith 1982). Some inventive graphical methods have been instigated that would be dull and unbearable to realize without computer support. The progress of *exploratory data analysis* (Cleveland 1985) has been determined in part by formation of innovative graphical tools for data depiction. With the advent of GIS technology, geographers now have a surfeit of graphic and cartographic practices (e.g., depict of temporal data, geographical flows, animated displays, and 3-D imaging techniques) at their disposal (Kraak 1988).

Visualization is an imperative section of any exertion to appreciate, investigate or elucidate the dissemination of a phenomenon on the surface of the earth. According to McCormick et al. (1987) "*Visualization is a method of computing. It transforms the symbolic into the geometric enabling researchers to observe their simulations and computations…It studies those mechanisms in humans and computers which allow them in concert to perceive, use and communicate visual information*". International Cartographic Association (ICA) Commission on Visualization and Virtual Environments stated that "*Geovisualization integrates approaches from visualization in scientific computing (ViSC), cartography image analysis, information visualization, exploratory data analysis (EDA), and geographic Information systems (GIS systems) to provide theory, methods and tools for visual exploration, analysis, synthesis, and presentation of geospatial data*". As the technology advances and expands; there is a propensity to inquire accordingly more intricate demands. Understanding the characteristics of spatial data can be protracted by technologies that permit us to see the concealed.

2.6.1 Objectives of Visualizations

International Cartographic Association's recognized four visualization goals i.e., exploration, analysis, synthesis, and presentation. Di Biase et al. (1992) designated two aspects of geovisualization—*visual thinking* which generates and infers graphic presentation, and *visual communication* which targets at allocating knowledge in an easy-to-read pictorial form. MacEachren and Ganter (1990) proposed a two-stage model for interacting with geovisualization—*seeing-that*, the analyst searches for patterns in visual input, defined as 1st stage model; and the 2nd stage called *reasoning-why* i.e., the confirmatory stage of scientific inquiry. However, another important driving force for geovisualization is the growth of the *information technology* and its growth into the protuberant medium to publicize geospatial data and map. To fit the full potential of geospatial data, gcovisualization apparatuses are prerequisite to adapt to their users.

2.6.2 Cartographic Visualization

Cartographic visualization continually played an imperative role in human antiquity. The earliest instance of cartographic visualization even dated back to the Stone Age with map-like wall canvases portraying the surroundings of our descendants. The data visualization began in the 2nd century C.E. with data organization into columns and rows and progressing to the early quantitative representation in the

17th century (Friendly 2008). Since then, the art and science of map making has progressed uninterruptedly until today. According to the Interaction Design Foundation, French Philosopher and mathematician René Descartes laid the ground work for Scotsman William Playfair. A number of methods and procedures espoused from cartography visualization and deliberated useful information in visualization process. Moreover, the use of modern visualization technology deals numerous new promises for geographic visualization or geovisualization tasks. Geovisualization especially deals with the geospatial data (i.e., data that containing geo-referencing), may aid to reconnoiter, apprehend and interconnect spatial phenomenon (Yasobant et al. 2015). Haining (1990) decayed geospatial data into an immaterial feature space and geographic space with two or three dimensions. The most usual visualization technique for geographically referenced data is cartographic presentation of particular form and onto which the data of interest are contrived at their conforming co-ordinates. Possibly the eldest and most regularly used technique for visualizing point data is plot the location of study subjects using their Cartesian co-ordinates (Van Long 2009).

Spatial features are characterized by their locations and attributes. To display spatial feature on a map, we use a map symbol to indicate the feature's location and a visual variable with the symbol to show the feature's attributed data. The visual variable for data display includes hue, value, chroma, size texture, shape and pattern. Though, point maps are the meekest method to visualize the disease events information if the locations of events are known. Nowadays, cartograms typically stab to preserve the users' mental map by possessing parallel shapes and by conserving the adjacencies between the portrayed areas. An example of a Kala-azar distribution cartogram in Bihar is shown in Fig. 2.3.

2.6.3 2D Visualization

The primary methods for visualizing two-dimensional (e.g., two pairs of cardinal direction) scalar fields are colour mapping and drawing contour lines. Descartes established a two-dimensional (2D) co-ordinate system for exhibiting values, which in the late 18th Century Playfair saw possible for graphical statement of quantitative data (Friendly 2008). In 2D visualization process, Cartographers classify maps by function and by symbolization. By functions map can be general reference or thematic. The general reference map is used for general purpose, whereas thematic map is to show the distribution pattern of a theme. By map symbol, maps can be qualitative (e.g., appropriate portraying qualitative data) or quantitative (e.g., appropriate for communicating quantitative data). The most wide spread approaches to epitomize categorical but also numerical data are ***choropleth maps***. A choropleth (derived from Greek *choros*—'place' and *plethos*—'magnitude') map practices the graphic variables pronouncing properties of color/texture to display possessions of non-overlapping areas, such as state, districts, or other small administrative units. Choropleth maps aids to represent one or two attributes of the

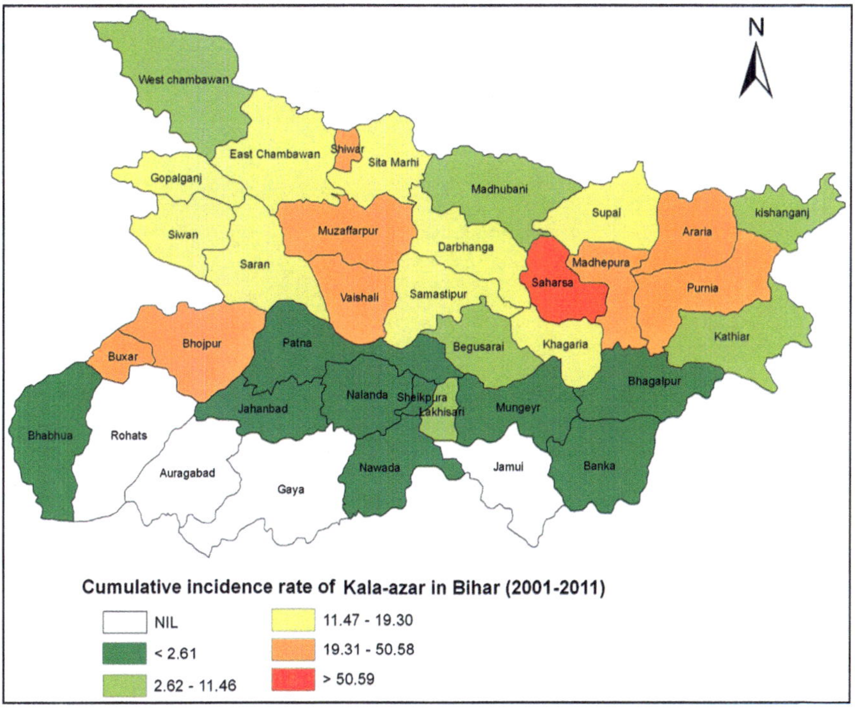

Fig. 2.3 In the cartogram, the cumulative incidence of Kala-azar distribution in Bihar is portrayed

data. Andrienko and Andrienko (1999a, b) designate selection procedures to epitomize single and many attributes in a map. Correspondingly, glyph-based methods from visual data mining can also be shared with the map presentations. Two instances of such methods are turn off faces (various variables of the data are connected to facial structures on an iconic face, such as shape and size of mouth, eyes etc.) and star plots (portray the value of set of attributes by the length of rays coming from the epicenter of the glyph). Two geometric procedures are frequently used in geovisualization- scatter plots (i.e., depict objects as points in a coordinate system where the axes correspond to two selected attributes) (Fig. 2.4) and parallel coordinate plots (i.e., displaying high-dimensional data in a single plot).

The ***dot map*** uses uniform point symbols to display spatial data, with each symbol demonstrating a unit value. One-to-one dot mapping uses the unit value of one, such as one dot demonstrating one case. But in most case, it is one-to-many dot mapping and the unit value is greater than one. For example, the spatial distribution of VL cases during the period between 1990 and 2012 was plotted though dot densities map (Fig. 2.5). The general distribution of Kala-azar cases in the study region exhibits an apparent clustering in the north-central part. Very few cases were found in the in the southern part of the study area. The results also illustrated several VL cases in the extreme north western and north-eastern part of the study

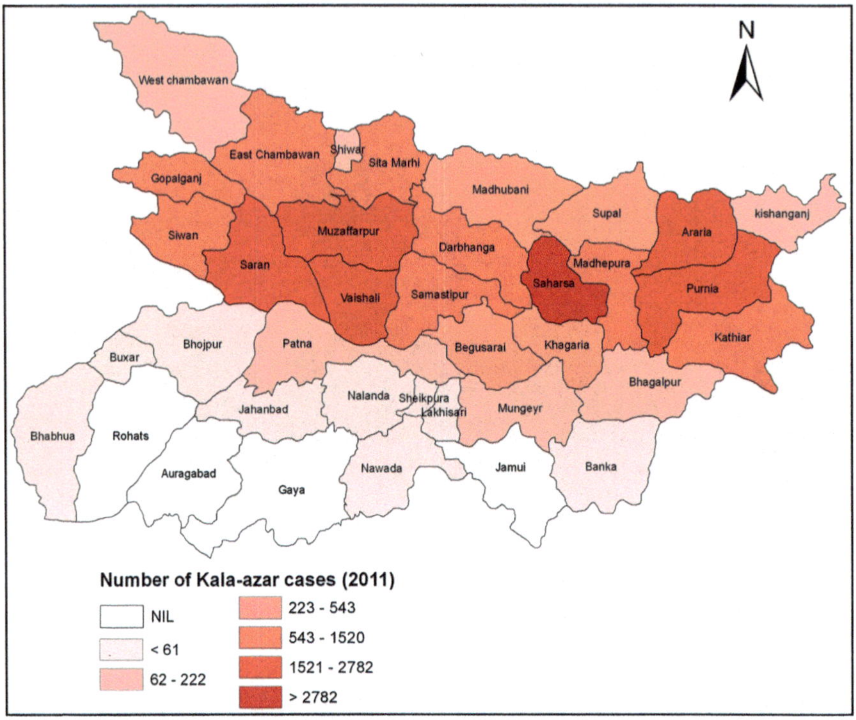

Fig. 2.4 A choropleth map showing the number Kala-zar distribution in 2011 in Bihar (India)

area. However, the northern plain region seemed to have greater amounts of VL infection than many other regions.

The ***graduated symbol map*** practices various-sized symbols such as circles, squares, or triangles to epitomize different ranges of values (Fig. 2.6). Two significant concerns to this map form are the range of sizes and the apparent dissimilarities between the sizes.

A ***proportional symbol map*** is a map that uses a symbol size for each numeric value rather than a range of values. Therefore, one circle size represents cases of 5000, 10,000 and so on.

The ***isarithmic map*** uses a system of isolines to represent a surface. Each isoline connects points of equal value.

The chart map practices either pie charts or bar/column charts. A disparity in graduated circle, the pie chart cans pageant two sets of quantitative data. The circle size can be prepared proportional to a value, such as cases, and the sub-divisions can display the makeup of the value (Fig. 2.7a). Bar charts use vertical bars and their summit to denote quantitative data. Bar charts are predominantly beneficial for likening data side by side (Fig. 2.7b).

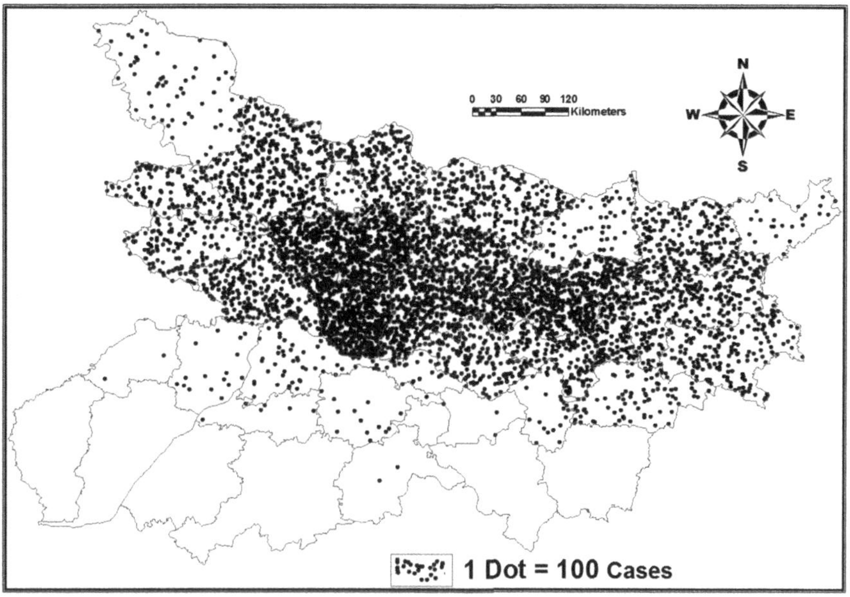

Fig. 2.5 Dot density map of total visceral leismaniasis cases in Bihar during the period between 1990 and 2012

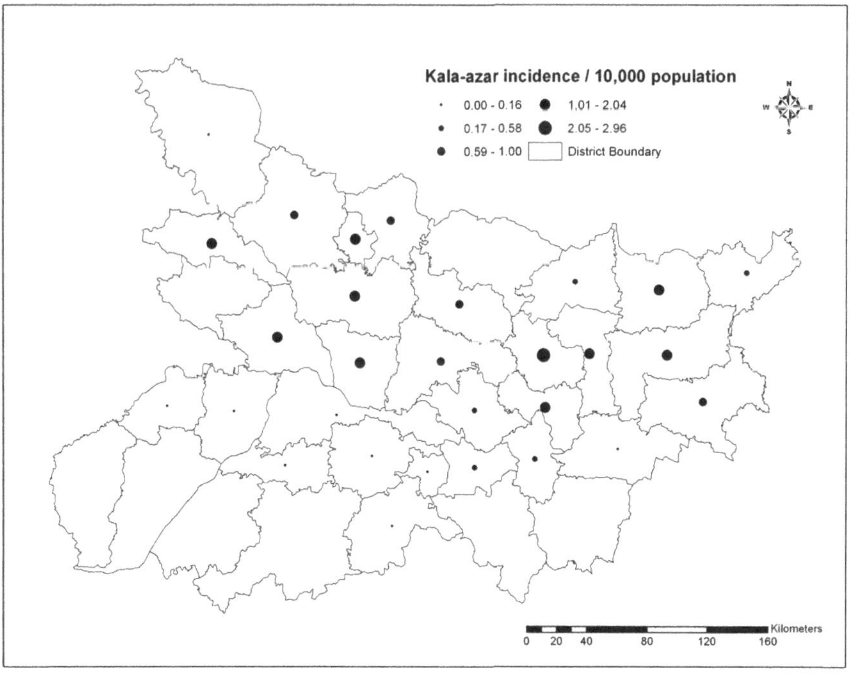

Fig. 2.6 Graduated symbol map of visceral leismaniasis (kala-zar) cases in Bihar in 2014

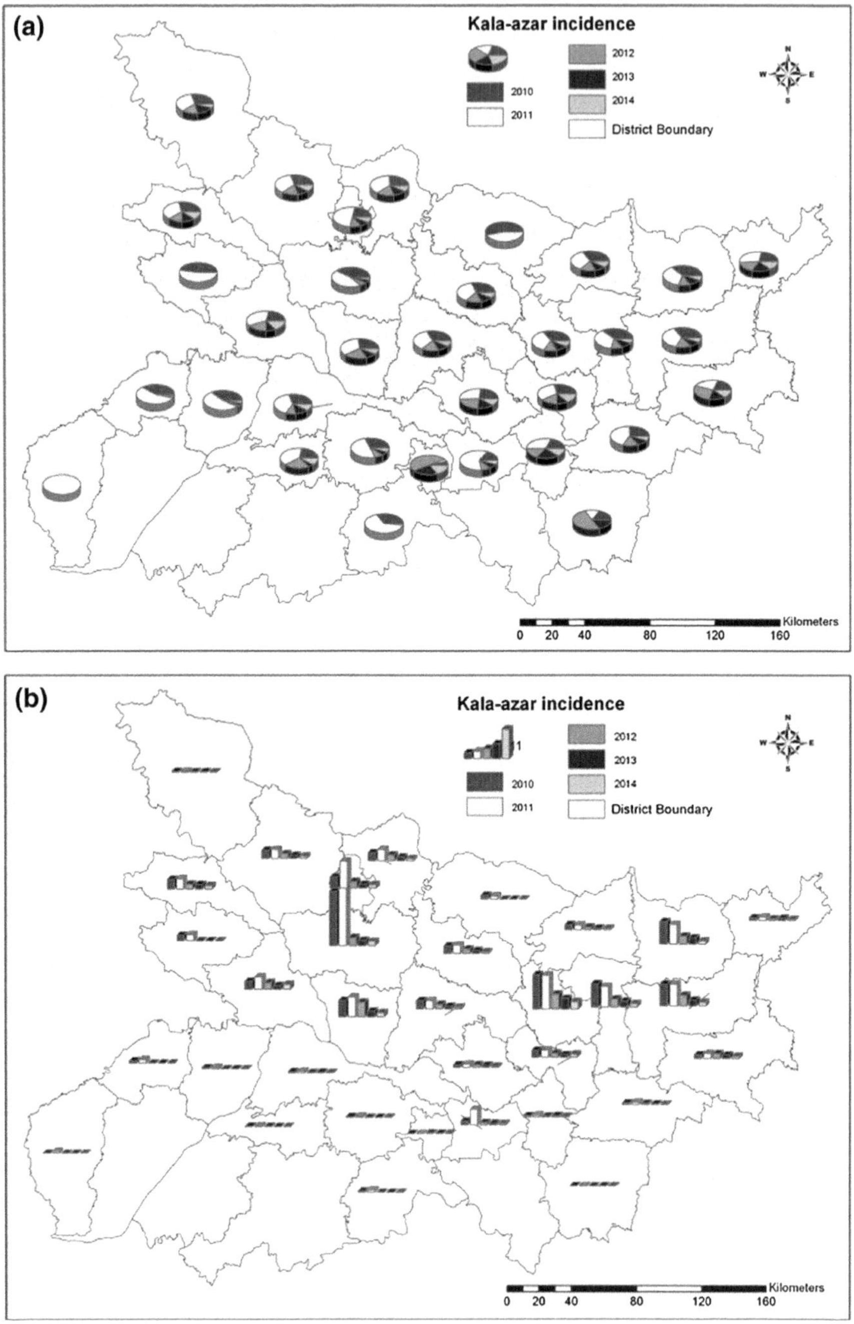

Fig. 2.7 Chart map of visceral leismaniasis (kala-zar) cases in Bihar during the period between 2010 and 2014 **a** representing pie chart map **b** representing bar graph

2.6.4 3D Visualization

3D visualization systems have been employed on visualizing orders that vary over time in a geo-spatial framework. Andrienko et al. (2003) deliberates numerous visualization methods for spatio-temporal data, with an emphasis on exploration. Kjellin et al. (2010 documented altered cases how 2D and 3D depictions outstrip each other. Cave Automatic Virtual Environments (CAVEs), and Power walls providing stereoscopic views to represent the 3D visualization of geospatial data. The 3D cartograms are based on the extrusion of the spatial data. There was used the ArcScene (part of ArcGIS software package) software application for statistical data representation using the 3D cartograms. ArcGlobe shows spatially referenced data on a 3D globe surface, exhibited in its true geodetic location. But MacEachren et al. (1997) stated that there have been only few efforts exploring abstract visualizations of geospatial data in 3D. However, the problem intrinsic in 3D visualizations more seems in the visualization of spatio-temporal data perhaps due to the upgraded dimensionality likened to chronological data. Alternative exertion with 3D visualization is difficult of creating precise size approximation because of the perspective foreshortening (Munzner 2008).

2.6.5 Spatio-Temporal Visualization

Spatio-temporal data are very common in public health to reveal, analyze and understand patterns of temporal changes of disease location and status. In case of spatio-temporal data, time can be perceived as a fourth dimension. The simplest form of this variety of cartographic visualization is the display of two or more maps, contiguous to one another, viewing the similar geographic area at various times. The change in the area portrayed is apparent in the difference(s) between various maps. Andrienko et al. (2000) classified spatio-temporal data in relation to the type of temporal variations: (i) existential variations (appearance and disappearance of features) (ii) alterations of spatial properties (variation of location, shape and size and (iii) vicissitudes of thematic possessions (qualitative and quantitative characteristics). All these practices tangled a map display to envisage the spatial attributes of the data. Shanbhag et al. (2005) presented three techniques that modify choropleth map to display the temporal attribute data—(i) clock like comparison for observing time and divisions the area polygon into wedges, (ii) representation of yearly rings of a tree trunk and allocates time points to rings of the polygon (iii) linear percentage time (e.g., polygons were separated into vertical pieces). Peuquet (1994) explicitly discriminates three constituents in spatio-temporal data: space (where), time (when) and objects (what). Based on that, the basic types of queries are promising:

- *when* + *where* → *what*: Define the location of entities at a given point of time.
- *when* + *what* → *where*: Designate the situation employed by a particular object at a given time.
- *where* + *what* → *when*: Determine the times of a particular entities employed a particular location.

2.6.6 Combination and Interaction of Visualization

Interaction is of paramount importance in geovisualization, particularly for graphical investigation. Interaction is crucial factor differentiating geovisualization from traditional geography (MacEachren 1995). Nevertheless, the manual interaction practices are time consuming and often provide only very inadequate discernment into the data. Buja et al. (1991) announced taxonomy for common interactive multivariate data visualization. A different method of interaction alarms the way in which the actual data items are displayed, i.e., deselected attributes as well as the actual numeric values of selected attributes should be open. However, the visualization and interaction methods designate so far which seem most efficacious for data exploration jobs in small and medium-scale datasets. As such computational approaches have been established in areas like machine learning and data mining that can examine enormous data measurements and mechanically excerpt knowledge.

2.6.7 Visualization Tool

Several visualization tools are available today to aid the users to decipher their respective tasks. As a substitute, tools need to be supple and certainly configurable by users in order to be pertinent to more than just a strongly demarcated problem. ArcView (http://www.esri.com/software/arcgis/arcgisonline) is a commercial tool for visualizing and examining geographic data and it compromises a number of methods for making customized thematic maps and analyzing spatial data. XGobi (http://stat-graphics.org/graphics/), primarily released in 1996, is a universal data visualization system supporting connected scatter-plot matrices and parallel co-ordinate plots and it can also connect to ArcView software. Cartographic data visualizer (https://www.visualrsoftware.com/cmp/data-visualization-demo/) assimilates the mapping and intellectual data visualization mechanisms into a single application. It deals dynamic choropleth map, population cartograms and statistical graphics. Common GIS (http://commongis.jrc.it), developed by FraunhoferIAIS, comprises a multitude of geovisualization techniques from dynamic choropleth map, optionally combined with bar plots and pie plots. GeoVISTA studio (http://www.geovista.in/) is an open source software, usually used for geovisualization and static application. Quantum GIS, (https://2ra5-downloads.phpnuke.org/en/c155067/quantum-gis-qgis) developed by the Open Source Geospatial Foundation (OSGeo) offers competences for geospatial data investigation, editing and visualization, and additional functionalities.

Geovisualization tools comprise of numerous interconnected displays, each portraying representation of the data under investigation to excite visual thinking. Users can take benefit of computationally rigorous tasks such as "draping" a perspective assessment over a surface or generating the impression of three dimensions on a 2Ddisplay by multifaceted execution and shielding procedures.

2.7 Conclusion

Disease maps play an important part in evocative spatial epidemiology. Maps are beneficial for numerous determinations such as documentation of areas with suspected increasing in risk, design of premise about disease etiology, and evaluating necessities for health resource allocation. The creation of striking and informative maps accompanies the proper analysis of spatial epidemiological data. Therefore, selection of appropriate technique, scale, symbology and concerned maps with other forms of data display allocating with statistical generalization are facets of visualization that should be prudently deliberated in order to attain these objectives.

References

Andrienko G, Andrienko N (1999a) Knowledge-based visualization to support spatial data mining. In: Proceedings of intelligent data analysis. Springer, Berlin, pp 149–160

Andrienko GL, Andrienko NV (1999b) Interactive maps for visual data exploration. Int J Geogr Inf Sci 13(4):355–374

Andrienko N, Andrienko G, Gatalsky P (2003) Exploratory spatio-temporal visualization: an analytical review. J Vis Lang Comput 14(6):503–541

Andrienko N, Andrienko G, Gatalsky P (2000) Visualization of spatio-temporal information in theinternet. In: Tjoa AM, Wagner RR, Al-Zobaidie A (eds) Proceedings 11th international workshop on database and expert systems applications, Greenwich, London, UK, 4–8 September, 2000. IEEE Computer Society, Los Alamitos, CA, pp 577–585

Buja A, McDonald JA, Michalak J, Stuetzle W (1991) Interactive data visualization using focusing and linking. In: Proceedings, visualization '91, IEEE conference on visualization, San Diego, CA, pp 156–163

Cleveland WS (1985) The elements of graphic data. Wadsworth Publishing Company

Di B, MacE David A, John K, Catherine R (1992) Animation and the role of map design in scientific visualization. Cartogr Geogr Inf Syst 19(4):201–214

Friendly M (2008) Milestone in the history of thematic cartography, statistical graphics, and data visualization. http://www.math.yorku.ca/SCS/Gallery/milestone/milestone.pdf

Groop RE, Smith RM (1982) Matrix line printer maps. Am Cartogr 9(1):19–24

Haining RP (1990) Spatial data analysis in the social and environmental sciences. Cambridge University Press

Kjellin A, Pettersson LW, Seipel S, Lind M (2010) Evaluating 2d and 3d visualizations of spatiotemporal information. ACM Transa Appl Percept (TAP) 7(3):19
Kraak, M-J (1988) Computer-assisted cartographical three-dimensional imaging techniques. Delft
Lehner B, Reidy Liermann C, Revenga C, Vörösmarty C, Fekete B, Crouzet P, Döll P, Endejan M, Frenken K, Magome J, Nilsson C, Robertson JC, Rodel R, Sindorf N, Wisser D (2011) High-resolution mapping of the world's reservoirs and dams for sustainable river-flow management. Front Ecol Environ 9(9):494–502
MacEachren AM (1995) How maps work: representation, visualization and design. Guilford Press
MacEachren AM, Ganter JH (1990) A pattern identification approach to cartographic visualization. Cartographica 27:64–81
MacEachren AM, Polsky C, Haug D, Brown D, Boscoe F, Beedasy J, Pickle L, Marrara M (1997) Visualizing spatial relationships among health, environmental, and demographic statistics: interface design issues. In: 18th international cartographic conference, Stockholm, 23–27, pp 880–887
McCormick SD, Saunders RL, Henderson EB, Harmon PR (1987) Photoperiod control of parr-smolt transformation in Atlantic salmon (Salmosalar): changes in salinity tolerance, gill Na+ , K+-ATPase and plasma thyroid hormones. Can J Fish Aquat Sci 44:1462–1468
Munzner T (2008) Process and pitfalls in writing information visualization research papers. In: Information visualization. Springer, pp 134–153
National Research Council—National Academy of Sciences (1995) National science education standard. National Academy Press, Washington DC
Nixon KMG (2011) A visualization of dissident voices in sixteenth-century Italy: a reflection of the religious debate in art PhD (Doctor of Philosophy) thesis, University of Iowa. http://ir.uiowa.edu/etd/1044
Open Geospatial Consortium (2011) OGC standards and cloud computing. Reference number of this OpenGIS© Project Document: OGC 11-036. file:///C:/Documents%20and%20-Settings/Administrator/My%20Documents/Downloads/OGC_Standards_and_Cloud_Computing_-_White_Paper_11-036.pdf
Peuquet DJ (1994) It's about time: a conceptual framework for the representation of temporal dynamics in geographic information systems. Ann Assoc Am Geogr 84(3):441–461
Ramankutty N, Foley JA (1999) Estimating historical changes in global land cover: Croplands from 1700 to 1992. Glob Biogeochem Cycles 13(4):997–1027
Sankoh O, Herbst A, Juvekar J, Tollman S, Byass P, Tanner M (2013) INDEPTH launches a data repository and INDEPTH Stats. Lancet Global Health 1:e69
Shanbhag P, Rheingans P, des Jardins M (2005) Temporal visualization of planning polygons for efficient partitioning of geo-spatial data. In: IEEE symposium on information visualization. INFOVIS. IEEE, 2005, pp 211–218
Shoultz J (2005) Enterprise real time geocoding for public health analysis and visualization. In: 2005 ESRI Health GIS conference paper 1060. South Carolina Department of Health and Environmental Control
Van Long T (2009) Visualizing high-density clusters in multidimensional data. PhD thesis, School of Engineering and Science. Jacobs University. http://vcgl.jacobs-university.de/wp-content/uploads/2011/06/Tran_Van_Long_PhD_thesis.pdf
Van Panhuis W, Paul P, Emerson C, Grefenstette J, Wilder R, Herbst AJ, Heymann D, Burke DS (2014) A systematic review of barriers to data sharing in public health. BMC Public Health 14:1144
Wartenberg D, Thompson WD (2010a) Privacy versus public health: the impact of current confidentiality rules. Am J Public Health 100(3):407–412
Wartenberg D, Thompson WD (2010b) Privacy versus public health: the impact of current confidentiality rules. Am J Public Health 100:40–412
Worboys MF, Duckham M (2004) GIS: a computing perspective, 2nd edn. CRC Press. ISBN 0415283752

World Health Organization (WHO) Technical Report Series (2010) Control of the leishmaniases. In: Report of a meeting of the WHO expert committee on the control of leishmaniases, Geneva, 22–26 March 2010. WHO technical report series; no. 949. ISBN 9789241209496. http://apps.who.int/iris/bitstream/10665/44412/1-/WHO_TRS_949_eng.pdf

Yasobant S, Vora KS, Hughes C, Upadhyay A, Mavalanka DV (2015) Geovisualization: a newer GIS technology for implementation research in health. J Geogr Inf Syst 7:20–28

Chapter 3
Basic of GIS and Spatio-Temporal Assessment of Health Events

3.1 Introduction: Geographic Information System

Geographic information system (GIS) is a computer based information system, consisting of computer hardware (CPU, Monitor, Keyboard, Mouse) and software (i.e., application software and programming software) and data (e.g., spatial and non-spatial) for capturing, analyze, arrange, manipulate to digitally represent and process geographic features present on the earth surface and linking the non-spatial attributes to it. However, GIS deals with the

- A map (spatial data) with a database (non-spatial data) behind it.
- A digital representation of the real world.
- Layers of information (Fig. 3.1).
- a class of software.
- an applied example of a GIS combines software with hardware, data, user etc., to decipher a problem, support a decision, help to plan.

Burrough (1986) defined GIS as "a set of tools for collecting storing, retrieving at will transforming, and displaying spatial data from real world for a particular set of purposes".

Aronoff (1989) defined GIS as a computer based system that provides four sets of capabilities to handle geo-referenced data.

GIS is not new-fangled. There have been four separate stages in the development of GIS.

Phase I: It has been started between the early 1960s and the mid-1970s, adage a new chastisement being conquered by a limited key individuals who were to form the way of future investigation and improvement.

Phase II: From the mid-1970s to early 1980s, saw the espousal of technologies by national agencies that ran to an emphasis on the development of best rehearsal.

Phase III: Between 1982 until the late 1980s, proverb the growth and misuse of the profitable market place adjoining GIS.

G. S. Bhunia and P. K. Shit, *Geospatial Analysis of Public Health*,
https://doi.org/10.1007/978-3-030-01680-7_3

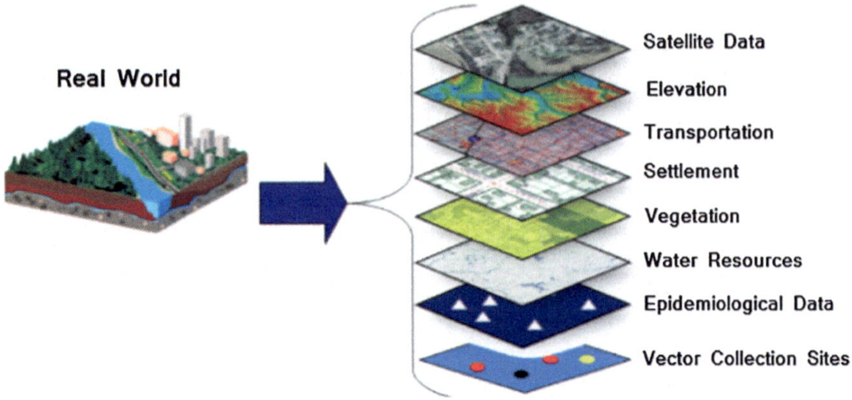

Fig. 3.1 Layers of information

Phase IV: Since the late 1980s has realized an emphasis on ways of enlightening the usability of technology by creating conveniences more users centric.

At the 20th Century, the Harvard Laboratory for Computer Graphics, the Canada Geographic Information System, the Environmental Systems Research Institute and the Experimental Cartography Unit in the UK were the major impacts in the field.

3.2 Basic GIS Operation: Vector and Raster GIS

Geographically referenced data i.e., data describes the location and attributes of spatial features, separate GIS from other information system. The data model states how geographically referenced features are epitomized in a GIS. Vector and raster data are the two main data forms used in GIS. Both vector and raster data have spatial referencing systems that determine locations on Earth.

3.2.1 Vector Data Model

In GIS, real-world objects are termed as spatial entities. A vector data model uses points and their x, y co-ordinates to construct spatial features of points, line and polygon (Fig. 3.2). Vector graphics are contained of vertices and paths.

- A ***point*** has zero-dimensions and has merely the possessions of location. A point may also be called a node, vertex or zero cells.
  Examples: wells, benchmarks etc.
- Lines typically characterize objects that are linear in nature. Cartographers can use a various breadth of line to display size of the feature/object. A ***line*** is

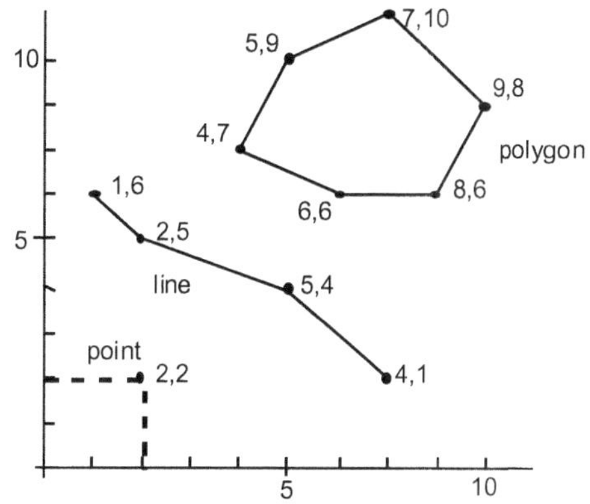

Fig. 3.2 Schematic diagram of vector data model

1-dimensional and has the characteristics of length. A line is also called edge, link, chain or 1-cell. It has the end points and shape of line may be a smooth curve or a connection of straight line segments.
Examples: Road, stream, canal etc.

- An ***area*** is 2-dimensional and has a size and perimeter (Chang 2006). An area is also called as polygon, face, zone or 2-cell.
  Examples: water bodies, land parcels etc.

Advantages: (i) Vector data are encompassed of paths. This means that graphical output is usually more aesthetically-pleasing. (ii) It provides greater geographic accuracy because data aren't reliant on grid size. (iii) Topology information can be kept with vector data models and permits for effective encoding of topology. (iv) Vector data structure is the model of excellent for competent network analysis and proximity maneuvers. (v) Data can be portrayed at its original tenacity and form without generalization.

Disadvantages: (i) Continuous data are poorly stored and exhibited as vectors. So to display continuous data as a vector would necessitate substantial simplification. (ii) Although topology is valuable for vector data, it is often processing intensive (iii) Location of each vertex essentials to be stored explicitly. For operational investigation, vector data must be transmuted into a topological structure. (iv) Spatial analysis and filtering within polygons are impossible.

3.2.2 Raster Data Model

Raster data model uses a grid and cells to represent the spatial variation of feature (Fig. 3.3). The data model stores images as rows and columns of numbers with a digital value/number (DN) for each cell. Units are usually represented as square grid cell that are uniform in size. The digital number in each grid cell resembles to the physiognomies of a spatial phenomenon at the cell location.

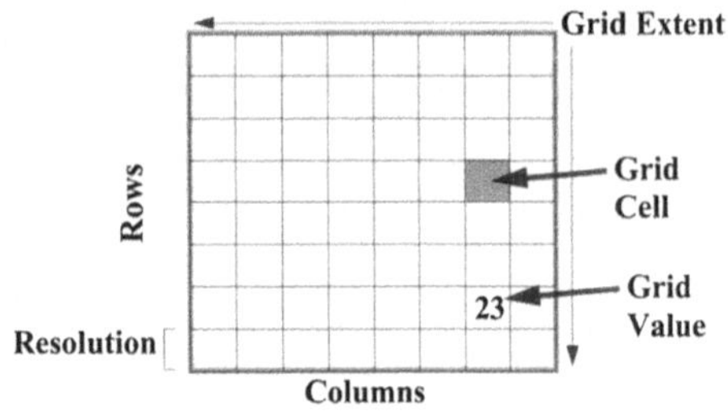

Fig. 3.3 Schematic diagram of raster data model

The origin of rows and column is typically at the upper-left corner of the raster. Rows function as y-coordinates and columns as x-coordinates. In raster data, point is represented in single cell, lines by sequence of neighboring cells and areas by collections of contiguous cells. However, the raster data model can be discrete or continuous.

- ***Discrete Raster***

Discrete raster is also called as thematic or categorical raster data, having definable boundaries. Discrete data usually consists of integers to represent classes. For example, value 1 represents the water body of land cover class and the value 2 represents the built up area.

- ***Continuous Raster***

Continuous raster is also called as non-discrete or surface data that denotes the grid cells with steady fluctuating data such as, elevation, temperature etc. A continuous raster surface can be consequent from *affixed registration point*. For example, a digital elevation model (DEM) is estimated from mean sea level. Each cell signifies a value above or below the sea level.

Advantage: (i) Raster grid format is the natural output of choice of satellite data. (ii) With cell size and a bottom-left coordinate, each cell location can be contingent. (iii) Data analysis with raster data is typically rapid and easy to accomplish. (iv) With map algebra, numerical analysis is spontaneous likewise the discrete or continuous raster.

Disadvantage: (i) Graphic output and quality is based on cell size. (ii) Linear features and paths are hard to demonstrate and depend on spatial resolution (iii) Networks are obstinate with raster data (iv) Raster datasets can become possibly of enormous size because a value must be documented and kept for each cell in an image. (v) Resolution increases as the size of the cell is reduced (Fig. 3.4).

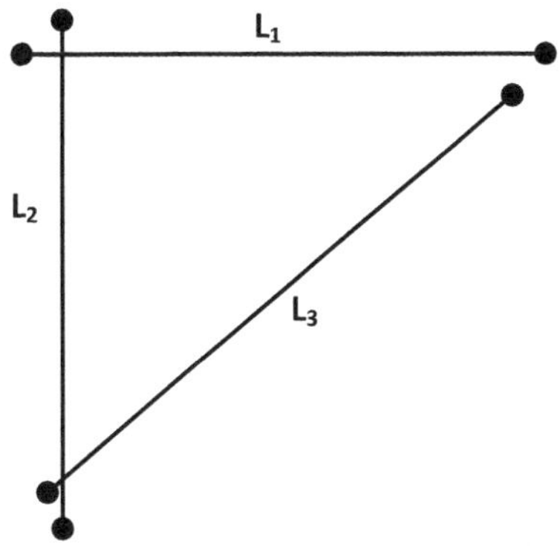

Fig. 3.4 Schematic diagram of Spaghetti data model

3.2.3 Spaghetti Data Model

The data model was one of the first conceptual models to edifice features in GIS. It was a simple GIS model where lines may be annoyed without intersecting or typology and typically no attributes is formed.

3.3 Spatial Analysis Within GIS

The origins of spatial analysis lie in the development of quantitative and statistical geography in the 1950s. Spatial analysis was originally based on the application of the available statistical methods to spatial data (Berry and Marble 1968). Later, it was extended to include mathematical model building and operational research methods (Taylor 1977; Wilson and Bennett 1985).

Hagerstrand (1973:69) gave a suitable explanation of spatial analysis when he marked "*to no small degree the recent quantitative analysis in geography represents a study in depth of the patterns of points, lines, areas and surfaces depicted on maps of some sort or defined by co-ordinates in two or three dimensional space*".

Gregory and Smith (1986:446) provide the definition of spatial analysis as "*quantitative procedures and techniques applied in locational analytic work*".

Goodchild (1988) stated that "*The true value of GIS lies in their ability to analyse spatial data using the techniques of spatial analysis. Spatial analysis provides the value-added products from existing datasets*".

Spatial analysis provides a toolbox that can in principle be applied to all the standard types of spatial information and can be accomplished in 1-d, more commonly in 2-d, occasionally in 3d, and rarely in 4-d space (Openshaw 1990). Spatial dependency and spatial heterogeneity are the crucial concept of understanding and examining spatial phenomena. Spatial dependence encounters with the typical hypothesis of independent observations in statistics. Conversely spatial heterogeneity is associated to spatial (or regional) distinction which follows from the intrinsic uniqueness of each location. Such idea stems from what Tobler (1970) calls the first law of geography: "*everything is related to everything else, but near things are more related than distant things*". Or, as ***Noel Cressie*** states that "*the (spatial) dependency is present in every direction and gets weaker the more the dispersion in the data localization increases*". Understanding the spatial distribution of data from phenomena that occurs in space creates today a pronounced contest to the explication of vital questions in many areas of knowledge, be it in health, in environment, in geology, in agronomy, among many others. These patterns allow the spatial conception of variables such as individual populations, quality of life directories or company sales in a region using maps. The emphasis of '***Spatial Analysis***' is to measure possessions and interactions, taking into account the spatial localization of the phenomenon under study in a direct way. To some extent the significance of spatial analysis to GIS is being documented and six crucial areas have been identified as follows:

- Response demonstrating for large data sets with diverse scales and measurement levels
- Applied methods from cross area calculation
- Zone order and spatial figure engineering
- Tentative geographical analysis technology
- Solicitation of Bayesian methods
- Application of artificial neural nets to spatial pattern detection.

3.3.1 Data Used in Spatial Analysis

The most used nomenclature to pronounce the hitches of spatial analysis cogitate three types of data:

- ***Events or point patterns***—Phenomena articulated through existences recognized as points in space, denominated point processes.
  Examples: disease occurrences, crime spots, and the localization of vegetal species.
- ***Continuous surfaces***—Appraised from a set of field samples that can be frequently or unevenly distributed. Generally, this type of data results from natural resources survey, like geological, topographical, ecological, phytogeographic, and pedological maps.
- ***Areas with counts and aggregated Rates***—Data related to population surveys, like census and health statistics, and that are initially denoted to individuals located in exact points in space.

3.3.2 Types of Spatial Analysis

The spatial analysis is composed by a number of connected measures whose goal is to pick out an inferential model that openly contemplates the spatial relationship existing in the phenomenon. There are several basic types of spatial analysis:

(i) *Queries and reasoning*: The most basic of exploration processes of GIS which is used to response simple queries pretended by the user. No modifications ensue in the database and no new data are created. Some of GIS software provide queries are as follows:

- *ArcGIS* offers full provision for customary SQL SELECT probes, such as: SELECT <all polygons in a given layer> WHERE <attribute table field> EQUALS <value>. It ropes JOIN and RELATE operations between attribute tables using definite fields, but it also delivers spatial additions to the JOIN and RELATE process allowing tables to be combined and queried based on their relative locations.
- The *ArcMAP* visual interface delivers a "Select by location" ability with combination capable by a change of spatial circumstances (appropriate to point, line and polygon features where evocative).
- *Manifold* offers sustenance for comprehensive spatial allowances to the standard SQL syntax. This allows a very large range of processes to be conducted, comprising so-called Action Queries (e.g. queries that CREATE, UPDATE and DELETE table data).
- *MapInfo* provides a sequence of SQL SELECT amenities and offers for spatial joins. 'Spatial join' operators buoyed include: 'Contains' (x comprises at least the other object, y's centroid), 'Contains Entire' (x encompasses entire object y), 'within' (as per Contain, but x and y characters are inverted), 'entirely within', and 'intersects' (x and y have as a minimum one point in shared). These operators allow attribute tables to be JOINed lacking of sharing a collective field (i.e., a spatial join only).

(ii) *Measurements*: Simple numerical values that define facets of geographic data. They contain dimension of basic properties of objects (e.g., length, area, or shape), and of the interactions between sets of objects (e.g., distance or direction). Practically all GIS packages deliver a varied range of measurement services. Arithmetic maneuvers and standard tasks beneficial to: (a) tabular data; (b) within queries and parallel tasks; and (c) smeared to map layers. A number of instances aid to elucidate the range of facilities provided:

- ArcGIS provides basic operations such as '+', '−', 'x' and '/' to be executed, with a range of standard functions such as cos(), exp(), log(), abs() etc. Operations are applied to one or more fields. Furthermore, an "Advanced" option is provided allowing Visual Basic for Application (VBA) instructions to be entered and performed. Such directions are often desirable in ArcGIS in order to add or update intrinsic data (e.g., the area of a polygon or the x, y coordinates of a polygon centroid).

- Manifold trappings such operations through its protracted SQL query amenities and by the Query toolbar ability it provides. The common arithmetic operators are employed and shared with comparison operators (e.g. = , >, < etc.) and Boolean operators (e.g. AND, OR, NOT, XOR).
- Many GIS packages permit the formation and saving of macros or scripts that tool frequently used map algebra expressions. For example, *TNTMips* and *ENVI* comprise many pre-built grouping formulas of this type. *TNTMips* holds such formulas as scripts that may be adapted.
- Within raster-based GIS, set of operations are provided (e.g., *IDRISI, PCRaster*), but are often supplementary apparatuses in common purpose (vector-oriented) GIS. For example, in '*ArcGIS*' the most of these tools are delivered in the 'Spatial Analyst' extension while in '*Manifold*' they are provided in the 'Surface Tools' extension.
- *PCRaster,* a non-commercial package intended explicitly for manipulating raster data, provides a strong set of point operations (Tomlin 1990). These contain a set of arithmetic, trigonometric and logarithmic functions, together with Boolean and relational operators.

(iii) *Transformations*: This performs to alter data sets by joining them or likening them to attain new data sets and finally new discernments. Transformations usage basic geometric, arithmetic, or logical rules and they comprise processes that transform raster data to vector data or vice versa. They may also generate fields from collections of substances or perceive gatherings of objects in fields. For example, its two-band transformed vegetation index (TVI) script is of the form:

$$C = 100(\sqrt{\frac{X-Y}{X+Y}}), X-Y \geq 0,\ X+Y \geq 0$$

$$C = 0\,if\,X - Y < 0$$

Here X might be the NIR (near infra-red) spectral band and Y the red spectral band, with the outcome being an index raster delivering a quantity of vegetation vigor.

(iv) *Descriptive summaries*: Experiments to seizure the soul of a data set in one or two numbers. They are the spatial correspondent of the descriptive statistics usually used in statistical analysis, comprising the mean and standard deviation.

(v) *Optimization techniques*: These are normative in nature; anticipated to choice seamless locations for objects' definite distinct circumstances. They are widely employed in souk exploration, in the suite dissemination industry, and in a mass of additional solicitations.

(vi) *Hypothesis testing*: Emphases on the progress of cognitive from the results of an imperfect sample to generate generalizations about a complete population. For example, to standardize whether a form of points could have soared by chance reliant on the substantiation from a sample. Hypothesis testing is the basis of inferential statistics and measures the principal of statistical analysis, but its practice with spatial data can be stimulating.

3.3.3 Common Error in Spatial Analysis

The spatial analysis is comprised by a cluster of chained measures whose aim is to designate an inferential model that evidently considers the spatial relationship existing in the phenomenon (Table 3.1). The vital problems in spatial analysis to numerous complications in analysis comprising bias, distortion and outright errors in the inferences reached.

- *Locational fallacy*

The locational fallacy refers to inaccuracy due to the specific spatial characterization taken for the features of study, particularly the choice of assignment for the spatial presence of the element. Spatial characterization may be simplistic or even wrong. Studies of humans often lessen the spatial animation of humans to a single point, for instance their home address. This can be easily lead poor analysis, for example when considering the disease transmission which can happen at work or at school and therefore far from the home.

- *Ecological fallacy*

The ecological fallacy describes errors because of performing studies on comprehensive data when annoying to find out conclusions on the individual units. An error happens in part of spatial accretion.

Table 3.1 Types of problems in spatial analysis

Type of analysis	Data types	Example	Typical problems
Point patterns	Localized events	Disease incidence	Determination of patterns and aggregation
Surface analysis	Samples of fields and matrixes	Vector distribution	Interpolation and uncertainty measure
Area analysis	Polygons and attributes	Epidemic data	Regression and distributions

- *Spatial uncertainty*

This type of error originates when objects do not have discrete or well demarcated extent or may have the effects that encompass beyond their boundaries.

- *Vagueness*

Vagueness occurs when the criteria that explain an object as '*x*' are not unambiguous or rigorous. For example, what incidence of malaria disease defines a high disease occurrence?

- *Ambiguity*

Ambiguity ensues when '*y*' is used as an auxiliary, or indicator, for '*x*' because '*x*' is not available. Undeniably, indicators are not simply direct or indirect; they conquer a scale. For example, relative humidity is direct indicator for larval growth of sandfly (visceral leishmaniasis). Altitude is an indirect indicator of sandfly species diversity.

3.3.4 Topological Error

Topological errors violate the topological relationship either required by a GIS package or defined by the user (Fig. 3.5). Topological errors with polygon features include unclosed polygons, gaps between polygons and overlapping polygons.

A common topological error with line features is that they do not meet perfectly at a point. This types of error are undershoot (if the gap exist between lines), overshoot (if a line is overextended), and pseudo nodes (divides continuous line unnecessarily).

Point features have few topological errors. It occurs when polygon has zero or multiple label points.

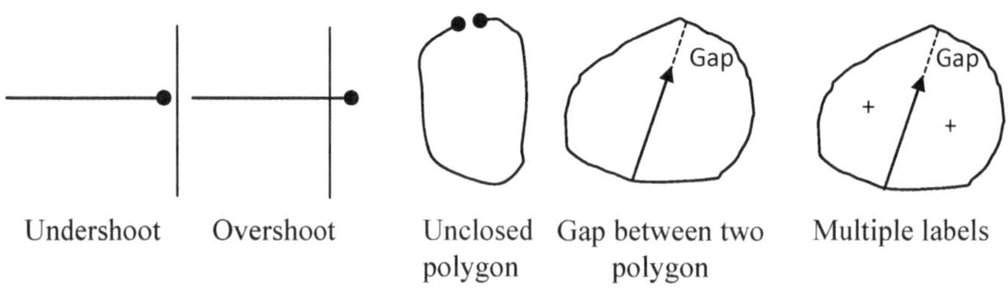

Fig. 3.5 Schematic diagram of various topological errors

3.4 Temporal Data Analysis and GIS

Temporal Geographic Information System (T-GIS) is an evolving competence in GIS for assimilating temporal data with place and aspatial data (Yuan 1996). The study of modeling temporal information in GIS began in the mid-1980s. Temporal data precisely states to times or dates, facilitating temporal imagining and eventually chronological analysis. Temporal data may denote to discrete measures.

Temporal GIS method accomplishes and studies spatio-temporal data (Nadi and Delavar 2003). In GIS, Langran and Chrisman (1988) pursuit the evidence of temporal GIS to sketch an outline for intangible strategy and application of integrating temporal information in GIS. Corresponding to the three relational database tactics, sequential information has been combined into GIS spatial data models by time-stamping layers (Armstrong 1988), attributes (Langran and Chrisman 1988) and spatial objects (Worboys 1992).

3.4.1 Time Series Analysis

To account for the developing physiognomies of observation data, time series analysis is a substitute for scrutinizing case incidences of health measures. The most systematic agenda uses time series models to predict probable numbers of cases, followed by assessment with the concrete observation. Appreciation of variations from historical events through envisages error use the distinction between the actual and predicted values at each point in time. In comparison to other monitoring system, time series methods use the relationship of the data at various time gaps in building estimates.

Consequently, time series analysis has previously been exposed to be somewhat valuable in diverse circumstances for supervising tasks. Its achievement into an assimilated scheme for use in health management will escort to an amended appraisal of its impact and usefulness. This will also provide the prospect for supplementary studies, such as the persuade of climatic and other ecological time series on the incidence of public health hazards.

3.4.2 Temporal Cluster Analysis

Identification of chronological clusters is implicit as amend in the incidence of disease event. This information is imperative to arouse investigation into the causes, and to hearten the progress of deterrent strategies. A substantial scheme for demarcating clusters is the scan statistic test (i.e., measuring the number of cases in all potential time intervals) (Naus 1965). Researches of sequential clusters based on the different time period between disease incidences have also been illustrated in the literature (Kanhabua et al. 2012; Meyer et al. 2015). These procedures presume that the random time intervals of consecutive cases form an autonomous and identically disseminated series of exponential unsystematic variables.

3.5 Spatio-Temporal Data Models and Methods

The emphasis of most GIS software is traditionally on the use geographical data rather than the spatio-temporal data. However, the vast quantities of spatial-temporal data now accessible hassles a re-think by numerous dealers of GIS and an allowance of their data models and systematic toolsets to hug these firsthand forms of data. One of the utmost imperative facets of this development is an alteration in perception, with the temporal field fetching ever more vital. Longley et al. (2010) provide a brief of spatial data models used in GIS and instance solicitations (Table 3.2). The differences are not as definite as they may perform. For example, transformation of vector data into raster and vice versa. Conversion in most cases will consequence in a forfeiture of information (e.g. resolution, topological structure) and thus such conversions may not be rescindable.

3.5.1 Spatio-Temporal (ST) Methods

(i) *T-mode data model*

It includes the spatial fields documented at discrete points in time, observed as a fixed time interval. The T-mode view of spatio-temporal field data is the most

Table 3.2 Geographic data models

Data model	Possible applications
Computer-aided design (CAD)	Automated engineering design and drafting
Graphical (non-topological)	Simple mapping
Image	Image processing and simple grid analysis
Raster/grid	Spatial analysis and modeling, especially in environmental and natural resource applications
Vector/geo-relational topological	Many operations on vector geometric features in cartography, socio-economic and resource analysis, and modeling
Network	Network analysis in transportation, hydrology and utilities
Triangulated irregular network (TIN)	Surface/terrain visualization
Object	Many operations on all types of entities (raster/vector/TIN etc.) in all types of application

common within GIS software. Analysis has inclined to emphasize on the variances between the time-sliced datasets.

(ii) *S-mode analysis*

It includes spatial fields noted at distinct points in time, observed as a regular of point locations or pixels, each of which has a temporal profile. This is extensively used in disciplines that study dynamic data over very enormous geographical extents. This represents spatio-temporal data as a form of space-time cube.

(iii) *Space and time analysis*

For such datasets study of the time-series data is often as significant as the procedure of approximating the whole spatial field. Partial spatial fields are measured at regular points in space and time. This type of data is usual of environmental monitoring using automated equipment like, weather stations, atmospheric pollution observing tools, river flow, and radiation monitoring devices.

(iv) *Time-stamping analysis*

Track data is normally a group of geographical coordinates together with a time-stamp for each coordinate. It is intended to be systematic then indiscretions designate that the spatial component of the track is erroneous during that period (i.e. the path from place X to the subsequent location detected Y, may be lost some intermediary points).

Example: velocity, acceleration and direction.

(v) *Network-based data*

Generally used for traffic checking, but event data on networks or connected to networks (e.g. crime events, accidents, transaction data, trip data, environmental monitoring data).

(vi) *Patterns of points over time*

This data is demonstrated by epidemiology, where embryonic designs of diseases (human, animal) are observed at a static location.

(vii) *Patterns of regions over time*

This type of analysis is mainly concerned with the census-based information but smears to an extensive variation of data which is composed on a systematic basis by region.

Example: socio-economic and health district data are typical examples.

3.5.2 Challenges of Spatio-Temporal (ST) Analysis

Data problems are an imperative aspect in spatial epidemiology (Teodoro et al. 2016). The investigation of spatio-temporal data is necessitated by both sequential correlations and spatial correlations of the data events. Evaluating both the temporal and spatial dimensions of data joins important intricacy to the data analysis procedure for two major reasons: (1) Continuous and discrete alterations of spatial and attribute possessions of spatio-temporal features and (2) the effect of collocated bordering spatio-temporal objects.

The modifiable areal unit problem refers those analytical results, using the same data for the same study area can differ extremely when amassed in different ways (Openshaw 1984). Most of the data concerned in spatial epidemiology depends on the data composed for other tenacities and necessitates gathering from numerous sources, in some cases, converting to a mutual boundaries.

Exposure misclassification arises when environmental planes do not precisely epitomize real exposure. Expectancy times between exposure and disease inception and population migration in and out of the area under study can source exposure misinterpretation.

3.5.3 Spatio-Temporal (ST) Data Analysis Workflow

The main goal of spatio-temporal analysis is to provide sufficient information to make out or to assess whether the data is appropriate and providing resources for further the analysis.

(i) *Collect and Prepare Data*

The important condition for spatio-temporal analysis is that all data must be connected equally to a spatial and a temporal constituent. Spatial data can be examined on several levels, geographical aspects (census tract, geocode etc.) or relative aspects. Temporal data is frequently investigated as various data points per observation over time (i.e., year, season, month, daily basis).

(ii) *Map and Examine*

Next step of spatio-temporal data collection is to examine the data for descriptive analysis and to scrutinize the data by simple descriptive maps. These tasks help to clear visualization of imperative appearances or tendencies that may be connected to spatial data that we may not realize by just seeing at the data.

(iii) *Pre-Process*

One more crucial aspect is to examine for non-independence of geographically associated interpretations. Prerequisite is to be anxious about clustering, and based on what types of data looks like and what type of clustering users are expecting.

(iv) *Define and Model Spatial Structure*

There are numerous models that are contained within the spatio-temporal agenda and that can be employed for these types of studies, like conditional auto-regression, space-time autoregressive integrated moving average, P-spline models etc.

(v) *Evaluate Model*

The appropriately developed model is measured by the common features of ST variation when the distributions for every aspect seem as random noise.

(vi) *Utilize Results*

Finally, the model has been suitably assembled, attuned and outcomes are ensured, and the results can be used in decision-making. Explanation of the consequences based on whether the model is developed to portray original patterns in health mapping or to outcome patterns of prospect disease.

3.6 Spatial Epidemiology

The leading accepted presentation of what could be assigned as a GIS was in France in 1832. An effort was carried out to extant and squabble worked from the data presented on the map. ***Charles Picquet*** (French Geographer) generated a map based illustration of cholera epidemiology in Paris with dissimilar halftone colour pitches (Picquet 1832). A similar state of affairs headed to ***John Snow*** portraying cholera deaths in London using points on a map in 1854 (Snow 1855). In the 1990s, one of the biggest GIS dealers issued *ArcView* which was a desktop elucidation for constructing mapping systems via a Windows based interface. In the early 20th century, photozincography was employed to isolate layer from a map. Finally, the key sectors such as government, non-government bodies and services seem to be emerging a tactic to partaking data and there is indication of substantial sharing through such platforms as ere has been important involvement of data sets across public podia.

In medical geography, GIS encompasses a number of key area including disease mapping, spatial exposure assessment, disease risk assessment and cluster detection. Spatial epidemiology is mainly concerned with the two fundamental questions:

- Where and when diseases tend to occur?
- Where does such pattern exist?

Spatial epidemiology provide the answers between place and health, especially it appraises how the physical and social environment outline the health and well-being of various personages (Cromley and McLafferty 2011). Spatial epidemiology is also fretful with the spatial scale at which the analysis accompanied (Diez Roux 2011). As pointed out by Root (2012), “the impact of neighbourhoods on health is uniquely geographic". Spatially accumulating data contribute to rise of modifiable areal unit problem (MAUP). It has thus become clear that it is progressively significant to demeanor analysis at a number of granularities of scale.

Based on the systematic collection of quantitative information, the aims of spatial analysis are:

- the careful and accurate description of events in geographical space (including the description of pattern);
- Systematic exploration of the pattern of events and the association between events in space in order to gain a better understanding of the processes that might be responsible for the observed distribution of events;
- Improving the ability to predict and control events occurring in geographical space (Haining 1994).

3.6.1 Why Disease Mapping Is Important?

Disease mapping is the leading step toward indulgence in the spatial characteristics of health-associated glitches because specific kinds of information are emphasized in maps. Distributions of disease can be portrayed via several cartographic techniques, such as points, lines, and polygon. Disease maps deliver an instant graphical summary of multifaceted spatial informations and may recognize elusive forms in the data that are wasted in tabular demonstrations. They are employed variously for evocative tenacities, to make hypotheses as to etiology, for observation to focus areas at seemingly high risk, and to support strategy development and resource distribution. They are also beneficial to support particular disease clusters and consequences of point-source trainings in appropriate circumstance.

Several disease mapping softwares that can be utilized in Epidemiology are as follows:

- SaTScan is a free software that analyzes spatial, temporal and space-time data using the spatial, temporal, or space-time scan statistics (http://www.satscan.org/download_satscan.html).
- GeoDa is a collaborating environment that pools maps with statistical charts and graphics, using the technology of animatedly linked windows. It was developed by Luc Anselin of the Spatial Analysis Laboratory of the University of Illinois, Urbana–Champaign (http://geodacenter.asu.edu).
- R has many spatial analysis packages, e.g. gstat package (http://gispopsci.org/spatial-analysis-in-r/).
- GRASS (Geographic Resources Analysis Support System) is free and open source GIS software set used for geospatial data management and analysis, image processing, graphics and maps production, spatial modeling, and visualization (https://grass.osgeo.org/download/).

3.6.2 *Epidemiology and Spatial Analysis*

Spatial epidemiology can be demarcated as the explanation and examination of geographic disparities in disease with respect to demographic, environmental, behavioral, socioeconomic, genetic, and infectious risk factors (Cromley and McLafferty 2011). Spatial epidemiology extends the rich tradition of ecology studies that use explanations of the distribution of diseases in different places to better understand the etiology of disease. GIS and spatial analysis provides unique tools to determine where and when particular disease has occurred and could resurface in the future. Accurate spatial data are thus critical to identifying such patterns.

Epidemiological data derive at various spatial scales (Deiz and Roux 2011). Root (2012) stated that "the impact of neighbourhood on health is uniquely geographic". Using dissimilar scales an investigation may clue to suggestively diverse outcomes. It has thus become clear that it is progressively imperative to demeanor analysis at numerous granularities of scale.

3.6.3 *Data for Spatial Epidemiological Studies*

Each item of health data (including population, environmental exposure, mortality and morbidity) may be connected with a point, or precise spatial position such as a home, a street address or an area, which could be defined as a spatial region by postcode, ward, local authority, province and country. A public health specialist may also come across spatial data in the form of continuous surface, such as the statistical surfaces of pollution interpolated from fixed-point characteristic (Rezaeian et al. 2007).

3.6.3.1 Point Patterns

Point patterns are based on the coordinates of events such as the positions of epidemics of a disease. It is also conceivable that they comprise attribute facts/figures such as the period of epidemic occurrence. The point pattern analysis will have to be perceived it is disseminated at random or characterizes a clustered or consistent pattern. It is imperative to know that the stochastic method deliberately narrates the positions where events are happening. First order properties of point pattern are enumerated in terms of the intensity of the procedure measured as the mean number of events per unit area. Second order properties describes the relationship between pairs of points or areas (i.e., analysis for clustering).

Point pattern analysis in spatial epidemiology concerns the distribution of disease events in space. At the elementary level, the spread of a disease in a community is revealed through the plotting of disease occurrences (at the residential locations of infected individuals) enabled with geocoding or address matching function in GIS. Point-by-point plotting is the simplest form of mapping disease occurrences (Lai et al. 2009).

The nature of pattern generated by biological processes can be affected by the scale at which the process is observed. Most of the natural environments show heterogeneity at a scale large enough to allow the emergence of patterns aggregates. On a smaller scale, the environmental variation may be less pronounced and the pattern will be determined by the intensity and the nature of the interactions between individuals.

Visualization of Spatial Point Patterns

The technique used most commonly to present spatial point patterns is a dot map. It is usually problematic to evaluate randomness of a pattern from visual inspection of such a map. Figure 3.6a, b present a series of dot map showing the location of Acute Encephalitis Syndromes (AES) cases in the Muzaffarpur district of Bihar (India) and the malaria incidences in Jharkhand district. Inspection of the map with spatial locations of AES cases could give the observer the impression that they are clustered in some particular locations. Consequently, Fig. 3.6b shows the intensity of malaria incidence comparatively high in the western part of the state. In both the cases it is found that diseases are not randomly distributed throughout the study region. But there is also some clustering of AES and malaria breakdowns.

Exploratory Analysis of Spatial Point Patterns

Methods for exploratory spatial analysis of point patterns are meant to originate instant statistics or plots of the experiential distribution to examine specific hypotheses. The approaches used are probing first or second order effects.

First order effects for point patterns can be examined with two techniques—*quadrat counts* and *kernel* estimation. The *quadrat* approaches involve dividing the area into sub-regions of equal size and provide a sign of the dissimilarity of the intensity of the essential process in space. *Kernel* estimation is a method which practices the actual point locations to create a suave bivariate histogram of intensity.

Second order properties of point patterns can be examined by the distances between the points—either the proximity between a randomly selected event and the nearest neighboring event or between a randomly selected location in space and the nearest event.

Modeling of Spatial Point Patterns

These methods are intended at elucidating an observed point pattern, and naturally comprise evaluation with the model of *complete spatial randomness* (CSR). A point

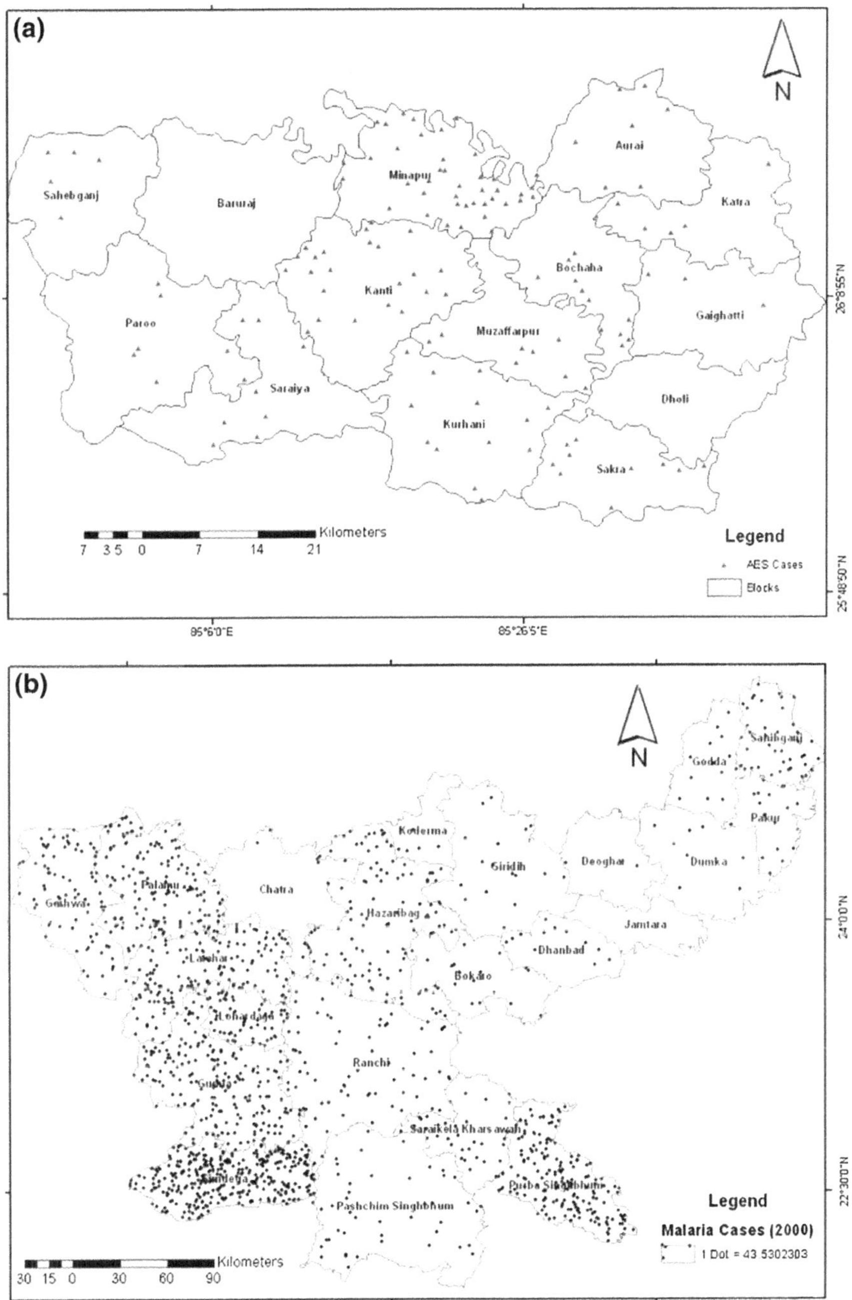

Fig. 3.6 Dot maps, showing the distribution of **a** Acute Encephalitis Syndromes (AES) of Muzaffarpur district (Bihar, India), **b** density of malaria incidence in Jharkhand (India)

pattern produced by a random spatial method should follow a *standardized Poisson process*. There are a range of systems obtainable to check for CSR. Some are based on quadrat counts such as the *index of dispersion tests*, others use *nearest-neighbor distances* such as the *Clark-Evans test* or the *K function*. Appraisal of an observed pattern with CSR has its precincts in epidemiology as it does not permit explanation of the type of point process other than whether it is entirely random in space or not. It also cannot proceed account of concerns such as a clustered essential population at risk. Other models which could be employed comprise the heterogeneous *Poisson process*, the *Cox process*, the *Poisson cluster process* or *Markov point process* (Bailey and Gatrell 1995).

Second order properties in a spatial procedure can be the consequence of disease clustering. Disease clustering can be measured using a lot of approaches and they can be characterized into general and focused tests (Waller and Lawson 1995). A tool which can be efficiently used for the investigation of grouping effects is the *K function* (Kingham et al. 1995). In this method, two groups of point practices such as cases of disease and random controls without the disease are likened. Bhunia et al. (2011) developed a method which is also based on nearest-neighbor distances. The test statistic simply compares second order nearest neighbour analysis to measure the distribution of disease incidence data locations according to whether they are clustered, random or regular.

3.6.3.2 Spatially Continuous Data

The point pattern analyses measured features of the spatial distributions of points, but makes only partial use of attribute information, also denoted as *geostatistical* data. The key role of the analysis will be to define the spatial variation in an attribute value, by the data collected at the sampled points.

Visualization of Spatially Continuous Data

The data values attained from the sampled locations can be plotted using evenly scaled symbols for each sampling point (Pfeiffer 1996). Overlapping symbols can generate a problem with explanation of such maps.

Exploratory Analysis of Spatially Continuous Data

The main techniques in this area are *spatial moving averages*, *tessellation methods* and *kernel estimation* techniques which are considered into the first order effect.

- A *spatial moving average* introduces values between a given numbers of adjoining sampling points. A weighting mechanism can be led to account for wavering distances between sampling points.

- *Tessellation* is most commonly done using *Delaunay triangulation*, also mentioned as a *triangulated irregular network* (TIN). This technique ascribes to each sampling point a territory in which each point is nearer to this sampling point than to any other. The resultant polygon map is called a *Dirichlet tessellation* and the tiles are called as *Voronoi* or *Thiessen polygons*. Such a TIN can be used to create a *contour map* (Fig. 3.7) or a *Digital Terrain Model* (DTM).
- *Kernel estimation* is used to transform the non-spatial data from the sampling points into a surface (Fig. 3.8). This method has been used in spatial epidemiology to model the relative risk purpose, calculating indigenous risk relative to the regional mean (Bithell 1990).
- The second order technique of exploratory analysis of spatially continuous data includes the *covariance* and *variogram* analysis. The *covariogram* defines the utility of the covariance for fluctuating distances *h* between sample points and the *correlogram,* the resultant correlation.
- The *semivariogram* is a graphical depiction of the variation between sampling points alienated by a specified distance and direction. Approximations of the *semivariogram* are measured to be more robust to partings from static characterized as a common trend in the spatial process. A stationary method will influence an upper bound, denoted to as the *sill* at a distance *h* termed the *range*. In theory, the intercept with the *y-axis* should be at a value of *0* variations. Practically, sampling error and small scale variation will consequence in inconsistency at small distances and the *variogram* will coincide the *y-axis* not in the origin. This intercept with the *y-axis* is called the *nugget* effect. Figure. 3.9 displays an isotropic sample *semivariogram* for the ratio of vector distribution captured at ground during the longitudinal study. The outline of the variogram proposes that the procedure is non-stationary, but specifies the comparatively small nugget value where there is the probability of spatial dependence.

Fig. 3.7 Examples of contour map and voronoi diagram representing spatially continuous data

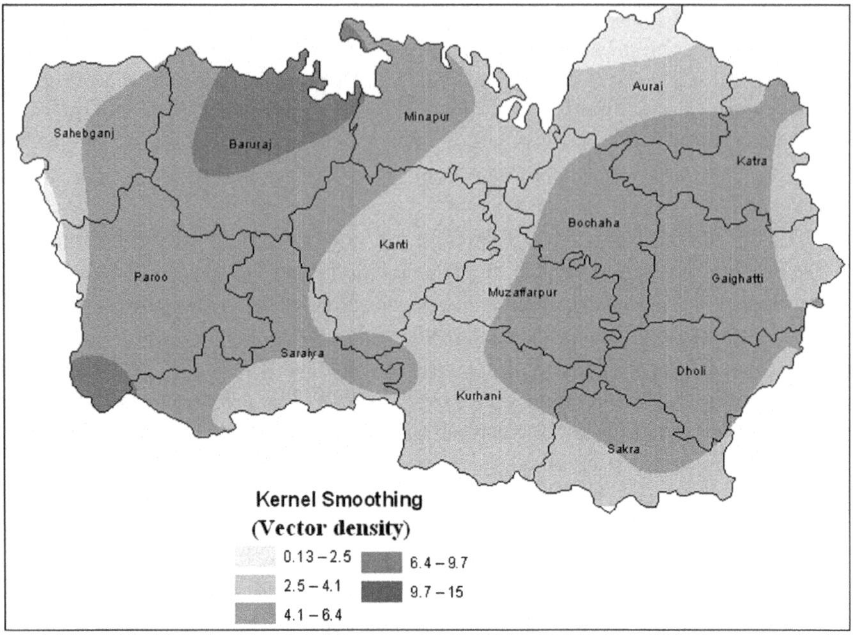

Fig. 3.8 Schematic diagram of kernel smoothing of vector density (*Phlebotomus argentipes*) of Muzaffarpur district of Bihar (India)

3.6.3.3 Area Data

Attribute data which does have values within fixed polygonal zones within a study area is referred to as area data or lattice data. The areal units can constitute a regular lattice or grid or consist of irregular units. The main emphasis with area data is on detection and explanation of spatial patterns or trends possibly extended to take account of covariates.

Visualization of Area Data

Area data can be envisaged by an extensive range of methods. The Choropleth map is perhaps the most usually suitable tool. Proper use of class interims and colors to epitomize values in a choropleth map is crucial. Cartograms or density leveled maps can be used to represent the significance of particular areas. The specialist has to be conscious of the difficulties which can be instigated by the changeable areal unit problem. It is also conceivable to show numbers of features at the definite time, by tallying scaled columns or symbols to a choropleth map (Fig. 3.10).

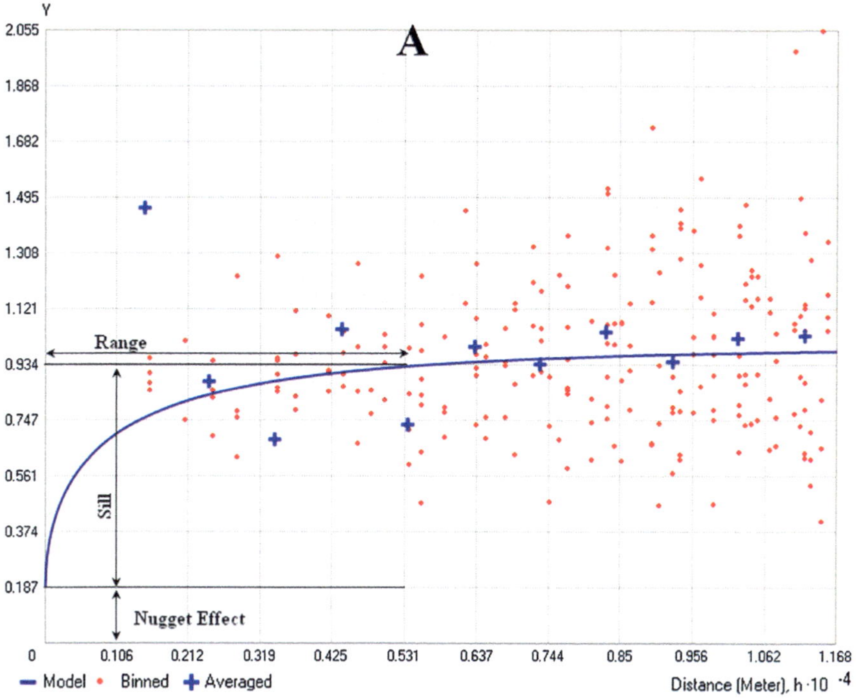

Fig. 3.9 Semi-variogram analysis of *Phlebotomus argentipes* (vector of visceral leishmaniasis) density of Muzaffarpur district during the period between 2011 and 2013 (Bihar, India)

Exploration of Area Data

Spatial weights matrix can be used to appraise spatial moving averages for investigation of first order effects. Kernel estimation can also be used for examinations of first order properties in area data.

In the case of second order spatial processes the objective is to explore spatial dependence of deviations in attribute values from their mean. The most commonly used techniques for spatial autocorrelation are *Moran's I* and *Geary's C*. *Moran's I* is closely associated to the covariogram and the *Geary's C* to the variogram used for geographically continuous data. A correlogram can be employed to present explicitly the correlation between values at various spatial lags. If the autocorrelation does not failure after a number of lags, it specifies the occurrence of non-stationary. The correlogram has parallel presentations in spatial analysis as it has in time-series analysis for describing patterns. Bhunia et al. (2013) analyzed the spatial distribution of visceral leishmaniasis cases between villages of Vaishali district in Bihar (India) using second-order analysis and identified significant spatial clustering within the district.

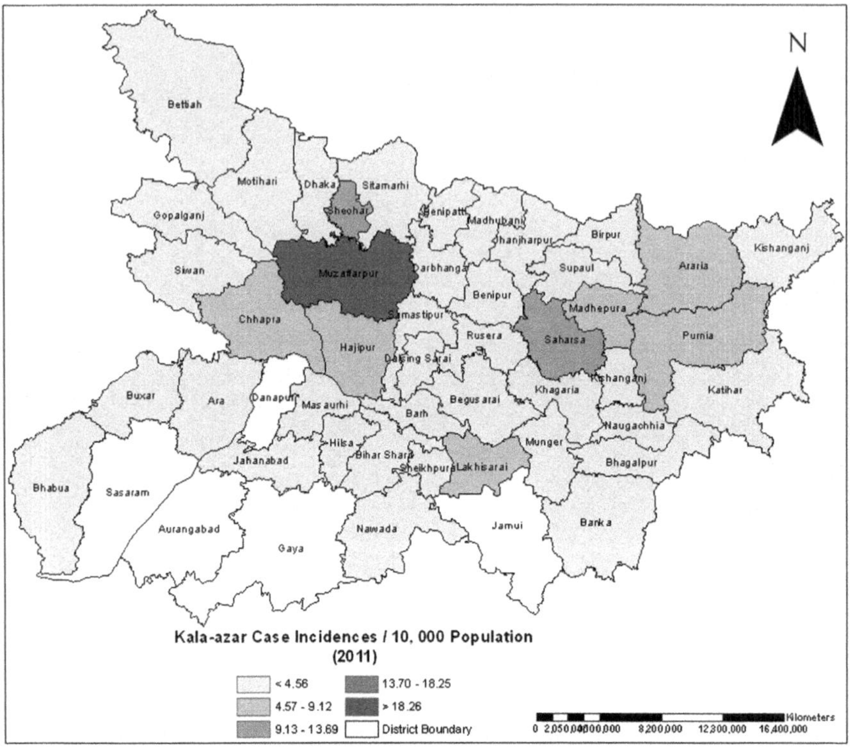

Fig. 3.10 Choropleth map of kala-azar (visceral leishmaniasis) case incidences per 10,000 population in Bihar (India), 2011

3.7 Case Study: Spatio-Temporal Distribution of Malaria in Jharkhand State (India)

3.7.1 Introduction

Malaria is an arthropod-borne infectious disease, executes a massive health and growing problem in South-East Asian region (Kumar et al. 2007). It is estimated that 90–160 million infections and more than 0.12 million deaths is taking place every year (WHO 2010). In India, the epidemiology of malaria is multifaceted because of geo-ecological assortment, multi-ethnicity, and outsized allocation of vector. However, in Jharkhand (extended between 25° 19′ 57.05″ N–21° 57′ 59.73″ N latitude and 83° 20′ 21.86″ E–87° 58′ 24.03″ E longitude), the situations of malaria is disquieting and contributing about 7% of total malaria cases in India. Malaria in Jharkhand has a prolonged and mutinous history. The total number of malaria cases in Jharkhand is taking the downward trend from 68,697 in 2000–37,482 in 2003. Since then, increasing trends were observed from 2005 with 43,388 cases, 2008 with 73,531 cases and 2009 with 91,194 cases (Source: National Vector Borne Disease Vector Control Programme and State Health Society, Jharkhand). The highest number of malaria deaths in Jharkhand was reported to be 61 in 2004 and this figure decreased to 22 in 2007. As per the report of National Vector Borne Disease Vector Control Programme (NVBDCP), disease incidences are on the upsurge with death information related with focal disease occurrences that are staying yearly distressing bordering population groups. Some initiatives have been undertaken by the Government of India to lessen malaria on the various regions where the disease still prevails (World Bank 2007).

Current tactics to improve understanding of the inconsistency in the epidemiology of the disease hinges on timely mapping and monitoring (Myers et al. 2009; Srivastava et al. 2009). However, the applicability of malaria risk mapping to formulate forecast at spatial scales and time points indispensable for effectual health service planning and evaluation depends principally on the amount and resolution of information offered (Saxena et al. 2009; Noor et al. 2009). Such vibrant maps based on current community-based data could be immensely treasured as a functioning utensil for planning programme against the designated malaria risk (Shirayama et al. 2009).

The present study aims to establish the changes in spatio-temporal distribution of malaria in Jharkhand from year 2000–2009 to reconnoiter the control of the interference coverage and the devotion to the intervention on malaria health outcome.

3.7.2 *Database and Methodology*

3.7.2.1 Study Area

Jharkhand, located in the northeastern part of the India, stretches between 1° 20′ N–6° 40′ N and from longitude 99° 35′ E–104° 20′ E. It covers 74,677 km^2 of geographical area (Fig. 3.11). The state comprised 32.97 million people as per Census data of 2011. The most prominent physical feature of Jharkhand is the Chota Nagpur plateau, part of the vast Deccan plateau that occupies most of peninsular India. In addition to the Damodar River in the northeast, the state is drained by the Subarnarekha River in the southeast and by the Brahmani River in the south. May, the hottest month, is characterized by daily high temperatures of 37 °C and lowest temperature of 20 °C. The rainfall ranges from a maximum of 1500 mm in the south-west and minimum of 1000 mm in the western-central part.

3.7.2.2 Study Design and Data Processing

Data were acquired for districts that were documented by administrative units. The study was carried out depending on retrospective secondary data of malaria for the

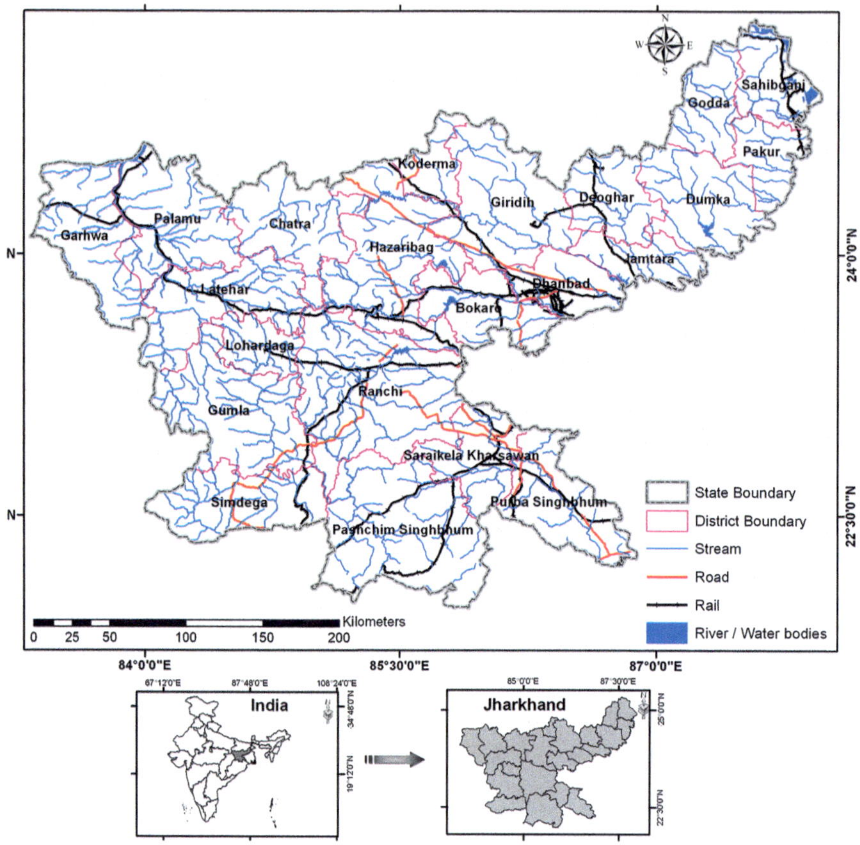

Fig. 3.11 Location map of the study area

period of 2000–2009. The incidences of malaria cases were combined at district levels and comprised information on malaria cases, type of parasites (*P. falciparum*, *P. vivax* and mixed infections), and time of illness (month and year). The annual reports of malaria, i.e., blood smears collected ("BSC"); blood smears examined ("BSE"); # of slides positive for *P. vivax* or malaria ("Pv"); # of slides positive for *P. falciparum* ("Pf"); and # of slides positive for both ("mixed") were calculated for the period between 2000 and 2009 from the Public Health Offices, Jharkhand. For the updated population numbers, the data were taken from the district Census report of Jharkhand, 2001 and 2011 which is available online. Decadal growth rate is used to calculate the mid-population of the study year. Collected data was then entered into the Excel file. After proper data cleaning, the data was analyzed using Microsoft Excel computer program and these raw numbers are then used to calculate several indices: percent of positive slides that are positive for *P. falciparum* ("%Pf" = (Pf + mixed)/positive); annual blood examination rate ("ABER" = BSE/

Population/10); annual parasite index ("API" = positive/Population); annual *falciparum* index ("AFI" = (Pf + mixed)/Population); slide positivity rate ("SPR" = 100 × positive/BSE); and slide *falciparum* rate ("SFR" = 100 × (Pf + mixed)/BSE).

3.7.2.3 Database Generation and Spatial Analysis

The spatial coordinates (the latitudes and longitudes) for each district were obtained from the DIVA-GIS (http://www.diva-gis.org/gdata). The annual malaria incidences in each district were mapped from 2000 to 2009 and portrayed to detect the annual fluctuations in Jharkhand state. Each district was marked with a various color on the district-level digital map. Summary Index of Malaria Surveillance (SIMS) was calculated for every year for each district based on principal components analysis (PCA) method followed by Cohen et al. (2010) to predict mortality. A digital database was generated to portray the spatial and temporal pattern of mortality at district level in the study site and used to reflect the extent of local malaria transmission. The software ArcGIS version 9.1 was used to describe the spatial distribution at the district level polygon map. The flow chart map of the study is represented in Fig. 3.12.

3.7.3 *Results and Discussion*

In Jharkhand, the disease distribution was spatially restricted in some pockets. Most deaths reported were inevitably happened of difficulties arising due to late reporting and resultant delayed treatment (source: State Health Centre).Spatial distribution of

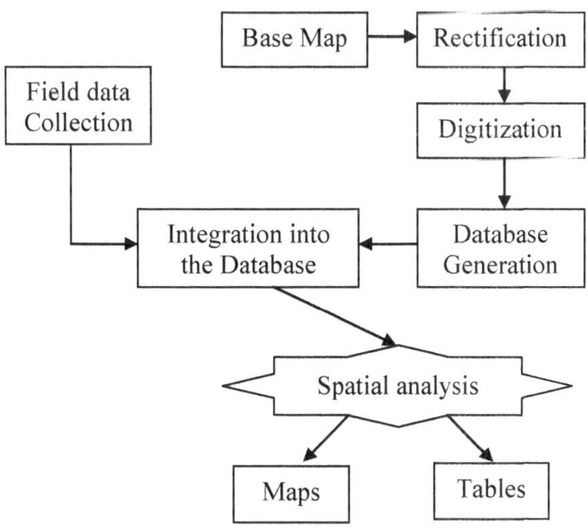

Fig. 3.12 Flow cart map of the study design

malaria incidences for a period of ten years in Jharkhand states is illustrated by digital mapping according to five parameters: ABER, API SPR, SFR and PF%. GIS map visually is indicated to present the uneven distribution of malaria incidences in the study area.

Descriptive statistics of annual blood examination rate (ABER) were calculated (Table 3.3). The highest average ABER was recorded in 2009 (12.27 per 10 inhabitants) and the lowest average of ABER was recorded in 2003. The results showed positive skewness, represents higher proportion of incidences relatively concentrated in some particular districts.

The variation of spatial distribution of malaria ABER for Jharkhand states during the period between 2000 and 2009 is graphically illustrated in Fig. 3.13. ABER reflects the efficiency and adequacy of case detection mechanisms. The areas were strategically colour coded based on the geometric interval as highest ABER (dark) and lowest (white). Within a ten year period, the values of ABER were constantly reported as more than 7.51% in west and south-west part of the state. Giridih, Deoghar and Jamtara reported the lowest ABER values of zero percent from 2000–2002. In 2003, Chatra, Ranchi, Paschimi Singhbhum, Purbi Singhbhum, Deoghar and Jamtara documented the lowest ABER values of zero percent. Table 3.3 represents the year-wise malaria ABER statistics.

The results of the analysis show that highest malaria ABER is recorded in 2009, while the lowest malaria ABER is recorded in 2003. By looking at the values, most of the districts in Jharkhand state still have the adequate and sufficient case detection mechanism.

Figure 3.14 shows the spatio-temporal distribution of malaria AFI in Jharkhand from 2000 to 2009. The central and eastern part of the state portrays the lower malaria AFI while the maximum value of malaria AFI is recorded in south-west part of the state and gradually it is shifted towards the southern part of the state.

Table 3.4 shows that highest value of AFI recorded in 2000 and the lowest value of AFI documented in 2007. The AFI values were reported >7.51 in Latehar,

Table 3.3 Descriptive statistics of annual blood examination rate (ABER) of Jharkhand State during the period from 2000 to 2009

Year	Minimum	Maximum	Mean	Stdv	Skewness	Kurtosis	Q_1	Median	Q_2
2000	0.54	12.33	4.26	3.31	1.12	3.16	2.33	2.88	6.1
2001	0.33	10.33	3.83	2.60	1.03	3.32	2.18	3.185	5.63
2002	0.32	9.83	3.57	2.30	1.01	3.72	2.18	2.77	4.77
2003	0.14	7.96	3.03	2.21	0.80	2.50	1.47	2.16	4.66
2004	1.39	16.87	5.57	3.99	1.34	4.15	2.70	4.43	7.12
2005	2.23	19.73	10.76	4.95	0.40	2.13	7.36	9.42	15.39
2006	2.96	14.00	7.88	2.98	0.32	2.33	5.94	7.00	9.35
2007	3.25	14.92	7.23	2.46	1.19	5.64	5.88	7.19	8.03
2008	0.00	12.60	3.24	3.72	1.50	4.11	0.61	2.40	4.15
2009	7.62	24.66	12.27	4.22	1.41	4.67	9.59	10.96	14.6

Stdv—Standard deviation; Q_1—Quartile$_1$; Q_2—Quartile$_2$

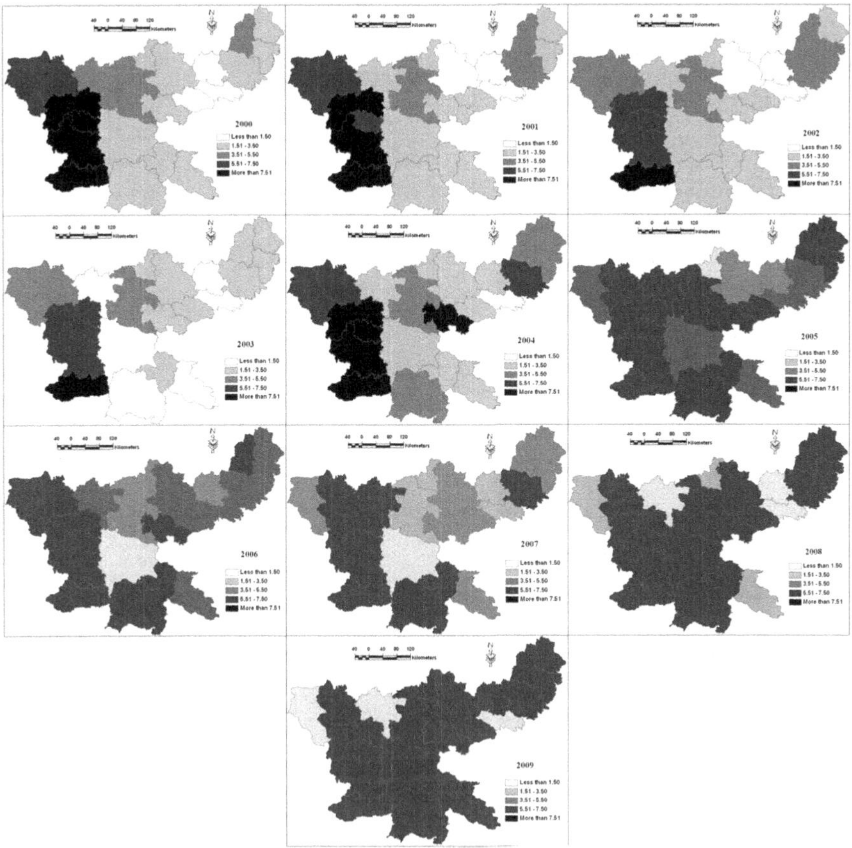

Fig. 3.13 The spatio-temporal distribution of malaria annual blood examination rate (ABER) in Jharkhand states from 2000 to 2009

Table 3.4 Descriptive statistics of annual *falciparum* index (AFI) of Jharkhand State during the period from 2000 to 2009

Year	Minimum	Maximum	Mean	Stdv	Skewness	Kurtosis	Q_1	Median	Q_2
2000	0.00	30.11	3.84	6.67	3.01	12.16	0.46	1.26	4.02
2001	0.00	25.60	3.41	5.71	2.91	11.57	0.35	1.61	3.87
2002	0.00	23.89	2.82	5.12	3.39	14.40	0.35	1.25	2.55
2003	0.00	18.45	2.05	3.90	3.59	15.71	0.23	0.61	2.22
2004	0.00	17.87	2.40	3.88	3.07	12.68	0.28	0.88	2.70
2005	0.01	21.02	2.58	4.70	3.06	12.14	0.26	0.82	2.43
2006	0.00	15.88	2.23	3.72	2.73	9.89	0.35	0.86	2.07
2007	0.00	12.68	2.01	3.05	2.44	8.47	0.25	0.89	3.00
2008	0.00	14.49	3.09	3.91	1.78	5.32	0.52	1.55	3.98
2009	0.01	16.16	4.02	4.73	1.28	3.54	0.68	1.6	6.13

Stdv—Standard deviation; Q_1—Quartile$_1$; Q_2—Quartile$_2$

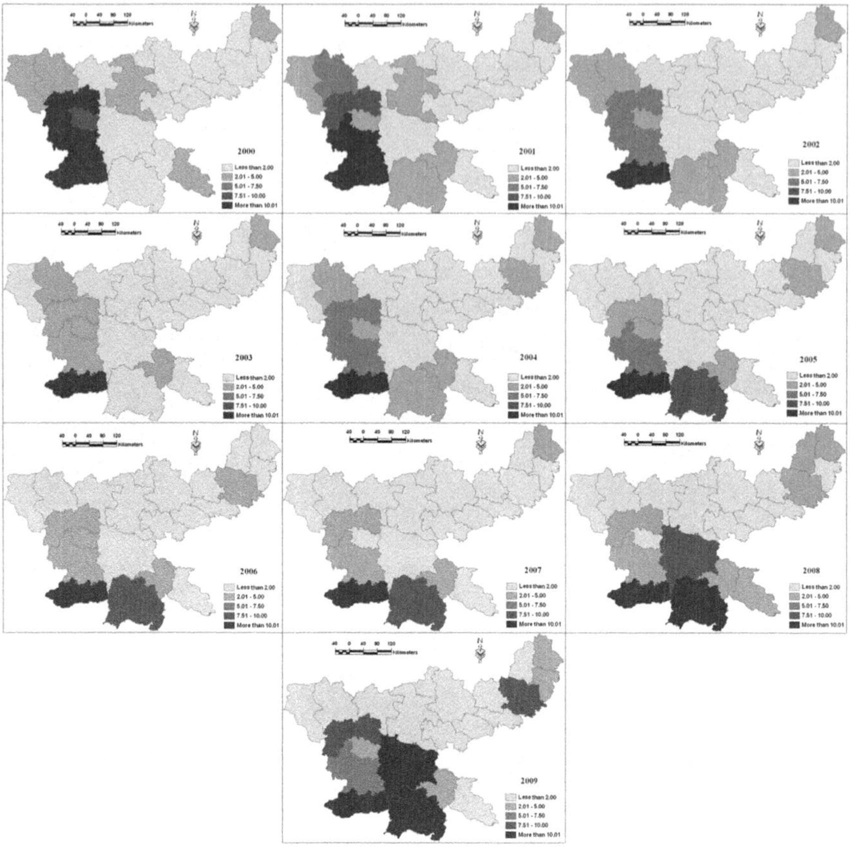

Fig. 3.14 Spatio temporal distributions AFI during the period between 2000 and 2009

Gumla, Lohardaga and Simdega during the period between 2000 and 2009. The central part of the study site stated the lower AFI values during the entire study period. Consequently, the result of our analysis portrayed gradual increase of AFI values in the north-east corner of the study site.

The temporal variations in API in Jharkhand state from 2000 to 2009 are shown in Fig. 3.15 and Table 3.5. API relies on the proficiency of case detection apparatus i.e., ABER. If ABER is ample, this parameter is the most imperative decisive factor to evaluate the advancement of the eradication programme. Deoghar, Jamtara and Dhanbad districts reported API values of <1.00/1000 population for a ten years period. Conversely, Garhwa, Palamu, Latehar and Simdega districts stated API values >15.00/1000 population. The central part of the state having medium to high API values during the study period. However, the API values gradually increase in the southern and north-west part of the study site. Therefore, to accomplish the goal, precedence and awareness in terms of malaria control activities must be given

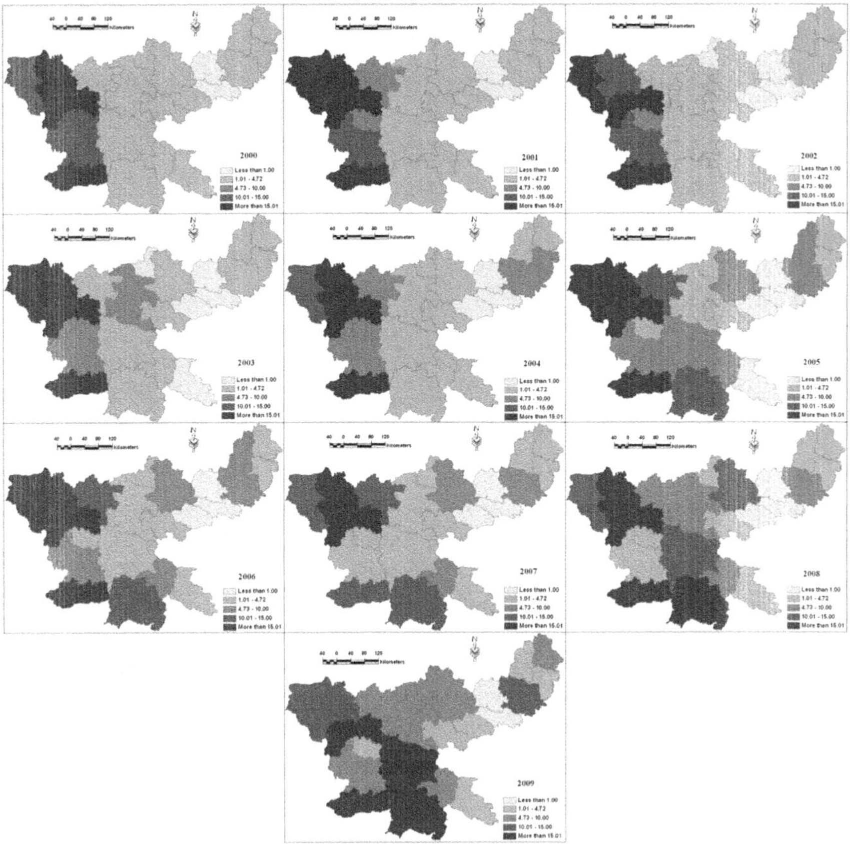

Fig. 3.15 Spatio temporal distribution of API during the period between 2000 and 2009

Table 3.5 Descriptive statistics of annual parasite index (API) of Jharkhand State during the period from 2000 to 2009

Year	Minimum	Maximum	Mean	Stdv	Skewness	Kurtosis	Q_1	Median	Q_2
2000	0.05	31.33	6.61	8.34	1.72	5.08	1.56	2.795	11.21
2001	0.02	26.42	6.47	7.71	1.48	3.57	1.65	2.66	7.44
2002	0.02	26.42	6.47	7.71	1.48	3.86	1.65	2.66	7.44
2003	0.00	19.36	5.2	6.18	1.44	3.48	1.03	2.84	5.89
2004	0.00	24.36	6.33	6.77	1.46	3.99	1.69	4.07	7.92
2005	0.03	26.63	7.96	7.64	1.17	3.28	3.09	6.21	11.24
2006	0.08	27.47	7.60	7.65	1.27	3.57	2.48	4.69	10.23
2007	0.01	29.67	7.09	7.46	1.60	5.11	1.98	4.05	11.80
2008	0.06	32.90	8.34	7.96	1.45	5.08	2.9	5.88	11.96
2009	0.02	41.04	9.34	9.34	1.92	7.06	3.47	6.5	14.04

Stdv—Standard deviation; Q_1—Quantile$_1$, Q_2—Quantile$_2$

Table 3.6 Descriptive statistics of percent of positive slides that are positive for *P. falciparum* ("%Pf") of Jharkhand State during the period from 2000 to 2009

Year	Minimum	Maximum	Mean	Stdv	Skewness	Kurtosis	Q_1	Median	Q_2
2000	0.00	96.11	47.04	27.34	0.18	2.04	27.85	47.305	67.63
2001	0.00	96.91	43.11	31.53	0.365	1.74	15.47	34.845	76.69
2002	0.00	96.46	42.32	30.55	0.29	1.76	14.12	38.67	69.8
2003	0.00	95.31	36.065	33.04	0.585	1.75	10.35	20.56	69.37
2004	0.00	96.51	36.29	30.21	0.43	1.896	9.03	32.575	67.43
2005	0.85	97.31	37.98	33.99	0.45	1.69	5.85	31.765	69.66
2006	0.00	97.04	36.69	31.14	0.40	1.81	4.35	28.935	64.72
2007	1.44	90.59	36.08	29.08	0.35	1.74	6.41	31.52	67.05
2008	0.00	110.89	46.73	32.64	0.02	1.93	14.56	49.585	73.97
2009	3.93	118.07	52.64	34.14	−0.04	1.92	22.43	57.97	79.72

Stdv—Standard deviation; Q_1—Quantile$_1$, Q_2—Quantile$_2$

to the high API value districts. A significant correlation was observed between ABER and API ($R^2 = 0.52$; $P < 0.019$) irrespective of the zones.

Percentage pf should be a good measure of the relative occurrence of *falciparum* and *non-falciparum* malaria in the study site (Fig. 3.15). The descriptive statistics of %pf is illustrated in Table 3.6. The result of our analysis shows that the lowest incidences of *P. falciparum* were observed in the northern and north-west part of the study site. The southern and north-east corner (West Singbhum, Simdega, Gumla and Khunti) was the worst affected recording highest %pf throughout the period. However, the %pf has clearly increased in the north-east corner (Sahibganj, Pakur, Godda, Dumka) during the entire study period compared to the previous years (Fig. 3.16).

Of total blood smears checked for malaria parasite, the prevalence rates for those positive for malaria and those positive for *P. falciparum* (SFR) were 3.58 and 2.28% respectively, but were noted to be variable among districts (Table 3.7). During the period between 2000 and 2003, SFR was higher (more than 7.5) in the western and southwest part of the state. However, the value of SFR gradually decreased between the periods of 2004–2007, and again increased in 2008–2009 (Fig. 3.17). The prevalence of *P. falciparum* over the study area did not show any uniform trend.

SPR measures the prevalence of malaria parasites among those who seek care and are examined in health facilities. Whenever ABER is adequate, SPR is a dependable parameter for determining the progress of measures and gives information of parasitic load in the community. Descriptive statistics of slide positivity rate (SPR) of Jharkhand state during the period from 2000 to 2009 is shown in Fig. 3.18 and Table 3.6. From 2004 to 2009, Deoghar, Bokaro, Dhanbad, Jamtara, Lohardaga and Gumla reported SPR values <5.0. SPR showed a declining trend over the years.

The result of our analysis derived from SIMS illustrated that frequency of morbidity in Jharkhand state gradually decreasing in trend during the period

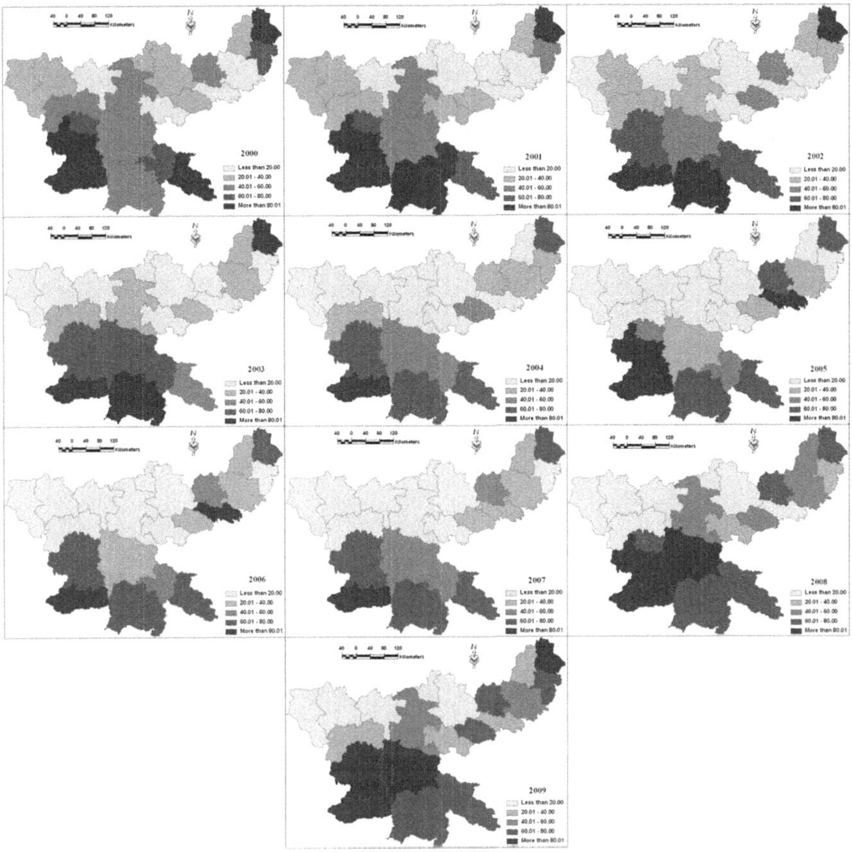

Fig. 3.16 Spatio temporal distribution of %pf during the period between 2000 and 2009

Table 3.7 Descriptive statistics of slide *falciparum* rate (SFR) of Jharkhand State during the period from 2000 to 2009

Year	Minimum	Maximum	Mean	Stdv	Skewness	Kurtosis	Q_1	Median	Q_2
2000	0.00	24.42	6.28	5.95	1.49	4.96	1.53	4.85	10.02
2001	0.00	24.78	6.028	5.83	1.48	5.79	1.26	5.38	9.07
2002	0.00	24.29	5.34	5.38	1.99	7.94	1.34	4.18	7.72
2003	0.00	23.17	4.77	5.44	1.91	6.95	1.1	3.09	6.92
2004	0.00	18.33	3.42	4.04	2.41	9.51	0.41	2.565	4.83
2005	0.04	13.66	2.04	3.21	2.66	9.53	0.39	1.055	2.24
2006	0.00	14.5	2.32	3.296	2.65	9.89	0.53	1.28	2.89
2007	0.01	12.22	2.44	3.22	1.97	6.12	0.47	1.13	4.05
2008	0.00	12.6	3.24	3.72	1.50	4.11	0.61	2.395	4.15
2009	0.01	12.3	3.07	3.43	1.39	3.97	0.64	1.56	4.22

Stdv—Standard deviation; Q_1—Quantile$_1$, Q_2—Quantile$_2$

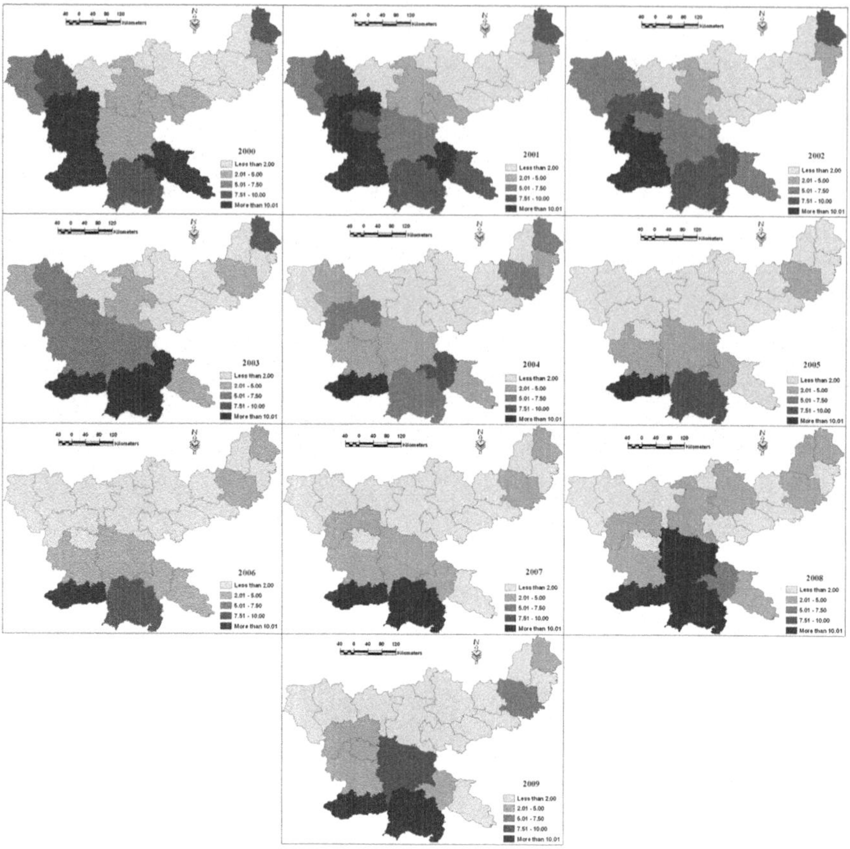

Fig. 3.17 Spatio-temporal distribution of SFR during the period between 2000 and 2009

between 2000 and 2007, and growing upward in 2008 and later (Fig. 3.19). The maximum frequency was estimated for the year 7of 2000, whereas the minimum value was recorded in 2007 (Table 3.8).

The SIMS shows a relationship as well or better than the entity measures with external measures of malaria mortality and morbidity of the study area. During the period between 2000 and 2002, the maximum frequencies of SIMS were concentrated in Gumla, Garhwa, West Singhbhum, East Singhbhum, Palamau district (Fig. 3.20). However, the concentration of the transmission intensity is slightly decreased in the northwest (Garhwa, Palamu, Latehar, Lohardaga district) corner of the study site. In Sahibganj district (extreme north-east corner) frequency of SIMS was maximum for the entire study period. Frequency of SIMS gradually decreased in the central part (Hazaribagh, Bokaro, Dhanbad and Giridih district) of the study area, however, the growing trend of SIMS's frequency was found in 2008 and 2009.

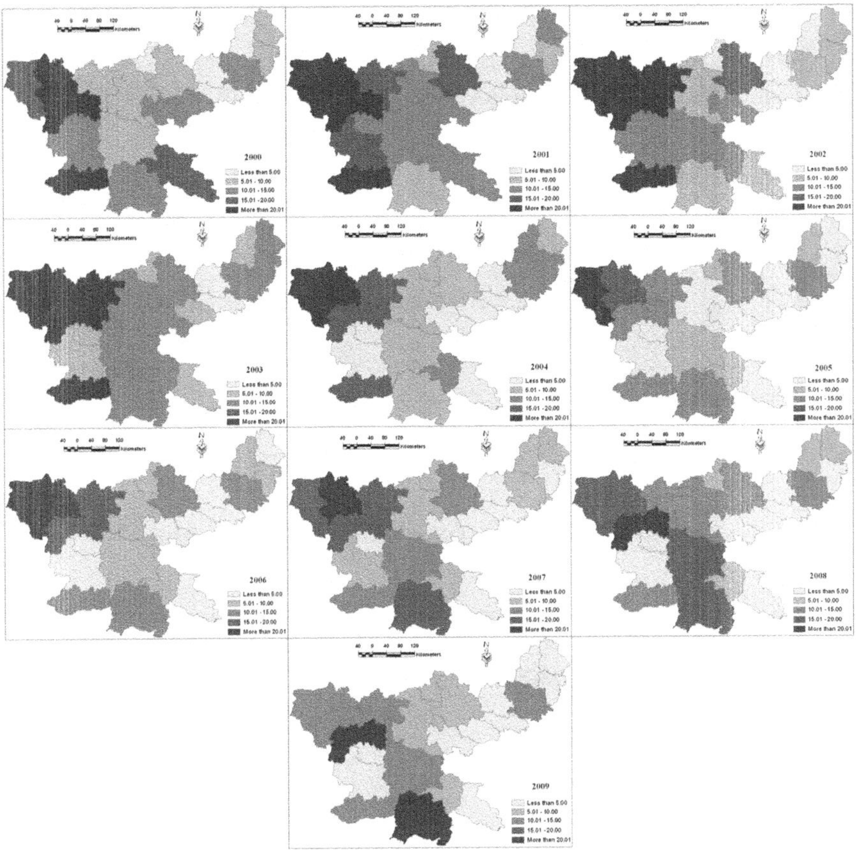

Fig. 3.18 Spatio temporal distribution of slide positivity rate (SPR) during the period between 2000 and 2009

The present analyses indicated that the Jharkhand state continued to remain highly malarious. The outcomes of this study proved that malaria endemicity is gloomier in the study area and *P. falciparm* was the predominant parasite species and its prevalence was high from the early part of the century. There was an intense variation in the spatial distribution of malaria within a ten-year period. Analysis on spatial distribution indicated that malaria problem was not homogenous in different parts of the state. Moreover, the SIMS described here should help the progress of the clarity of malaria surveillance. Therefore the feasibility of malaria control by application of a uniform strategy throughout the district is debatable. Moreover, several factors, such as profusion of vector, human activities, population immunity, social and economic status, and control measures, are known to have major

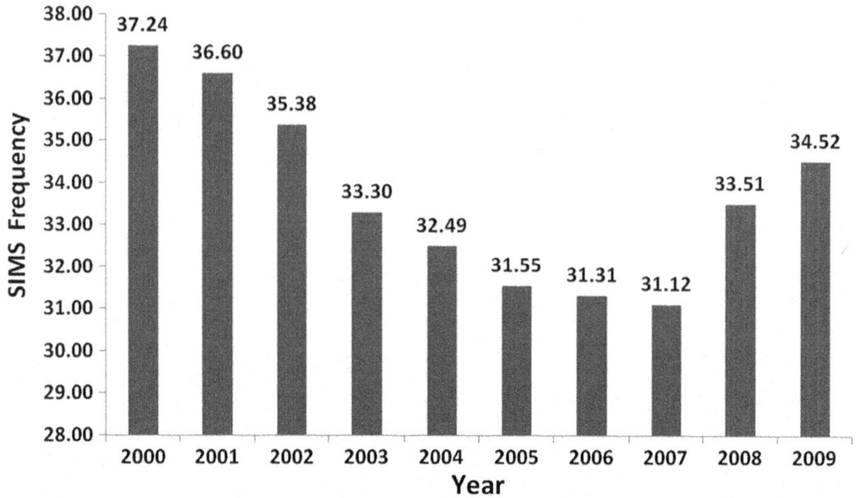

Fig. 3.19 Year-wise trend of SIMS during the period between 2000 and 2009

Table 3.8 Descriptive statistics of slide positivity rate (SPR) of Jharkhand State during the period from 2000 to 2009

Year	Minimum	Maximum	Mean	Stdv	Skewness	Kurtosis	Q_1	Median	Q_2
2000	0.00	30.11	3.84	6.67	3.01	12.16	0.46	1.26	4.02
2001	0.00	25.6	3.41	5.71	2.91	11.57	0.35	1.61	3.87
2002	0.00	23.89	2.82	5.12	3.39	14.40	0.35	1.25	2.55
2003	0.00	18.45	2.05	3.90	3.59	15.71	0.23	0.61	2.22
2004	0.00	17.87	2.40	3.88	3.07	12.68	0.28	0.88	2.70
2005	0.01	21.02	2.58	4.70	3.06	12.14	0.26	0.82	2.43
2006	0.00	15.88	2.23	3.72	2.73	9.89	0.35	0.86	2.07
2007	0.00	12.68	2.01	3.05	2.44	8.47	0.25	0.89	3.00
2008	0.00	14.49	3.09	3.91	1.78	5.32	0.52	1.55	3.98
2009	0.01	16.16	4.02	4.73	1.28	3.54	0.68	1.6	6.13

Stdv—Standard deviation; Q_1—Quantile$_1$, Q_2—Quantile$_2$

influence on the transmission of malaria. This is palpable from the digital mapping of the infection in Jharkhand. Outcomes of the study lay a substance for malaria observation and control in this area. In future, it will be beneficial to reconnoiter the method for conveying the association with the environmental factors to expand the predictive validity. Finally, it is recommended that the present findings should be deemed in potential malaria control and management programme and in the advancement of spatio-temporal models for transmission of malaria in Jharkhand state (India).

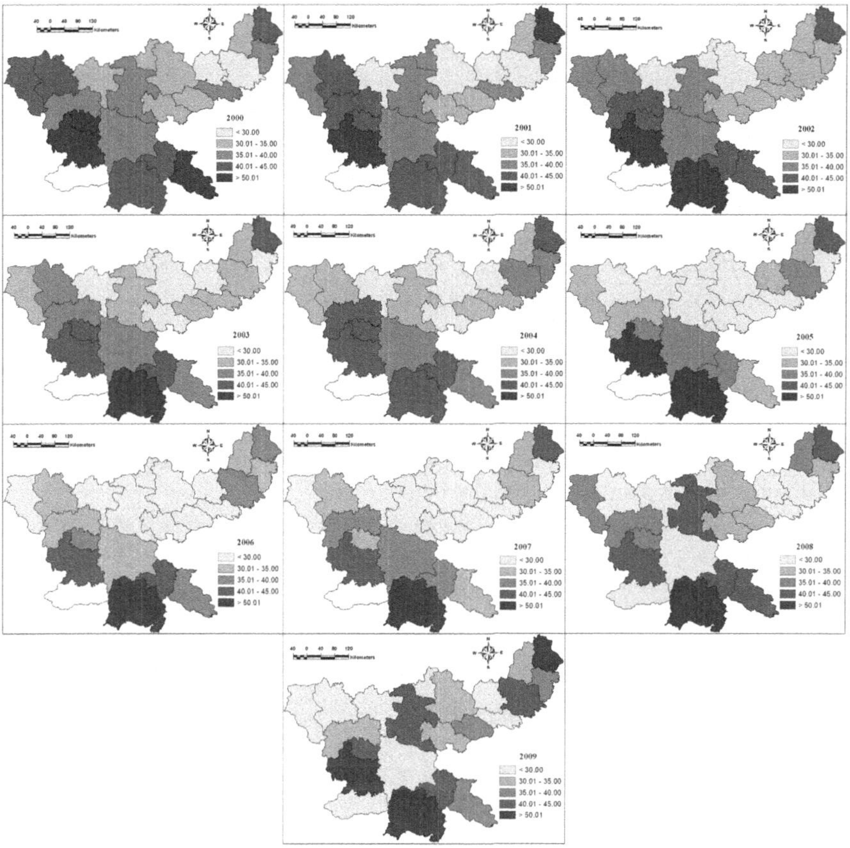

Fig. 3.20 Spatio-temporal distribution of SIMS during the period between 2000 and 2009

3.8 Benefits of Spatial and Temporal Analysis in Epidemiology

Disease distribution: The investigation of the transmission of communicable diseases (malaria, leishmaniasis, dengue, and AIDS) would help to form the assignment of disease incidence in a spatial-temporal ecological structure by GIS.

Easy perspective: Spatio-temporal mapping techniques assist to display quickly information and generate maps to highlight disease risk prone areas for developing more effective control and prevention strategies.

Handling enormous spatial data sets: As geographically referenced data sets become larger. This technology aids to develop methods to use, pare, classify, and manipulate the rich information inherent in very large spatial data bases.

Analyze spatial and space-time data: Exploratory methods of evaluating spatial data need to be extended to contain space-time data so that the model is to be developed to represent reality accurately.

Develop assenting measures: Statistical procedures must be developed to recognize the necessity of georeferenced data and permit investigators to involve in the testing of hypotheses.

Extend use of geo-statistical procedures: Geo-statistical measures are now being assimilated into GIS frameworks to models of continuous spatial phenomena.

Analyse the influence of scale and the growth of scale-independent measures: Feasibly the most important to organize GISs for spatial analytical work is the modules and algorithms that appraise the possessions that variations in scale have on research results. Investigators have great interest in the degree to which inter-zonal investigation is precious by the outline and regionalization of spatial data units.

Study global versus local effects: Procedures and assessments will be developed to use window, kernel, individual, and other local events to bargain and associate the features of non-stationary spatial data.

Improve computationally rigorous measures: Computationally intensive tools (neural nets, fuzzy sets, real-time data analysis) can permit more real practice of large data sets, and widespread imitations of multifaceted spatial phenomena, and the elucidation of compound location and distribution problems should be evaluated in GIS environment.

Use econometric modeling in a GIS environment: Spatial econometrics is a new and burgeoning field that can be employed to find suitable estimators and challenging manoeuvres for heterogeneous, non-uniform geographically referenced data.

Generate spatial interface models in a GIS framework: Evolving spatial interface models in a GIS environment will measure envisaging spatial interaction flows in a cultured manner (marketing, transportation, and human interaction).

Accommodate operations research: The operational study would profit prominently from the functionality and power of GIS to operate data, particularly in answering difficulties such as routing, location allocation, coverage, and other optimization.

3.9 Conclusion

Spatial data have become an important component of disease investigations. The availability of geographic information systems in combination with fast and relatively inexpensive computer hardware leaves the epidemiologist with the responsibility of making effective use of the information. The data necessities for a eloquent model with public health usefulness are reliant on first determining the concentrating of the exhibiting implementation, e.g., hypothesis analysis, recognizing knowledge breaches, providing way for investigation and control determinations, or assessing usefulness of an interference. GIS software and new mapping software nowadays deliver ability to produce risk maps in a multiplicity of designs comprising overlays on satellite imagery and vibrant artworks of space-time patterns. This is escorted by volatile growth in the field of web-based information conveyance which currently offers an operative mediocre to allocate risk maps to an extensive array of stake holders containing the medical community, vector control specialists, policy architects and the public at large.

References

Armstrong MP (1988) Temporality in spatial databases. Proc GIS/LIS'88 2:880–889

Aronoff S (1989) Geographic information system. A Management Perspecive, WDL, Ottawa

Bailey TC, Gatrell AC (1995) Interactive spatial data analysis. Longman Group, Harlow, Essex, England, p 413

Berry BJL, Marble DF (eds) (1968) Spatial analysis: a reader in statistical geography. Prentice-Hall, Eng, ewood Cliffs New Jersey

Bhunia GS, Kesari S, Chatterjee N, Kumar V, Das P (2013) Spatial and temporal variation and hotspot detection of kala-azar disease in Vaishali district (Bihar), India. BMC Infect Dis 13(1):1–13

Bhunia GS, Kesari S, Chatterjee N, Pal DK, Kumar V, Ranjan A, Das P (2011) Incidence of visceral leishmaniasis in the Vaishali district of Bihar, India: spatial patterns and role of inland water bodies. Geospatial health 5(2):205–215

Bithell JF (1990) An application of density estimation to geographical epidemiology. Stat Med 9:691–701

Burrough PA (1986) Principles of geographical information systems for land resources assessement. Clarendon, Oxford

Chang KT (2006) Introduction to geographic information systems, 3rd edn, McGraw-Hill, p 432

Cohen MJ et al (2010) Research identification of complex metabolic states in critically injured patients using bioinformatic cluster analysis. Crit Care 14(1):R10

Cromley E, McLafferty S (2011) GIS and public health. Guilford Press, New York

Diez Roux AV (2011) Investigating neighbourhood and area effects on health. Am J Public Health 91(11):1783–1789

Goodchild MF (1988) A spatial analytical perspective on geographical information systems (PDF). Int J Geogr Inf Syst 1:327–344

Gregory J, Smith E (1986) Procedimentos quantitativos (principalmente estatísticos) e técnicas aplicadas no trabalho analítico de localizações, p 446

Hagerstrand T (1973) The domain of human geography. In: Chorley RJ (ed) Directions in geography. Methuen, London, pp 67–87

Haining R (1994) Designing spatial data analysis modules for geographical information systems. In: Fotheringham AS, Rogerson P (eds) Spatial analysis and GIS. Taylor and Francis, London

Kanhabua N, Stewart A, Nejd W, Romano S (2012) Supporting temporal analytics for health-related events in microblogs. In: CIKM'12, 29 Oct–2 Nov, Maui, HI, USA

Kingham SP, Gatrel AC, Rowlingson B (1995) Testing for clustering of health events within a geographical information system framework. Environ Plann A 27:809–821

Kumar A, Valecha N, Jain T, Dash AP (2007) Burden of malaria in India: retrospective and prospective view. Am J Trop Med Hyg 77(6):69–78

Lai PC, So FM, Chan KW (eds) (2009) Spatial epidemiological approaches in disease mapping and analysis. CRC Press Taylor and Francis Group, Boca Ratón, FL

Langran G, Chrisman NR (1988) A framework for temporal geographic information. Cartographica 25(3):1–14

Longley PA, Goodchild MF, Maguire DJ, Rhind DW (2010) Geographic information systems and science. Third Edition. Hoboken, NJ: Wiley

Meyer S, Held L, Höhle M (2015) Spatio-temporal analysis of epidemic phenomena using the R package surveillance. https://arxiv.org/pdf/1411.0416.pdf

Myers WP, Myers AP, Cox-Singh J, Lau HC, Mokuai B, Malley R (2009) Micro-geographic risk factors for malarial infection. Malar J 8:27

Nadi S, Delavar MR (2003) Spatio-Temporal modeling of dynamic phenomena in GIS. In: ScanGIS 2003 Proceeding, pp 215–225

Naus J (1965) The distribution of the size of maximum cluster of points on the line. J Am Stat Assoc 60:532–538

Noor AM, Gething PW, Alegana VA, Patil AP, Hay SI, Muchiri E, Juma E, Snow RW (2009) The risks of malaria infection in Kenya in 2009. BMC Infect Dis 9:180

Openshaw S (1990) Towards a spatial analysis research strategy for the regional Research Laboratory initiative. In: Masser J, Blackmore MJ (eds) Geographical information management: methodology and applications. Longman, London

Openshaw S (1984) The modifiable areal unit problem. Norwich, UK: Geo Books

Pfeiffer DU (1996) Issues related to handling of spatial data. In: McKenzie J (ed) Proceedings of the epidemiology and state veterinary programmes. New Zealand Veterinary Association/ Australian Veterinary Association Second Pan Pacific Veterinary Conference, Christchurch, 23–28 June, pp 83–105

Picquet C (1832) Rapport sur la Marche et les Effets du Choléra-Morbus dans Paris et les Communes Rurales du Departement de la Seine. Le Ministre du Commerce et des Travaux Public, Paris

Rezaeian M, Dunn G, St Leger S, Appleby L (2007) Geographical epidemiology, spatial analysis and geographical information systems: a multidisciplinary glossary. J Epidemiol Community Health 61:98–102

Root ED (2012) Moving Neighborhoods and Health Research Forward: Using Geographic Methods to Examine the Role of Spatial Scale in Neighborhood Effects on Health. Ann Assoc Am Geogr 102(5):986–995

Saxena R, Nagpal BN, Srivastava A, Gupta SK, Dash AP (2009) Application of spatial technology in malaria research & control: some new insights. Indian J Med Res 130:125–132

Shirayama Y, Phompida S, Shibuya K (2009) Geographic information system (GIS) maps and malaria control monitoring: intervention coverage and health outcome in distal villages of Khammouane province. Laos Malar J 8:217

Snow J (1855) On the mode of communication of cholera

Srivastava A, Nagpal BN, Joshi PL, Paliwal JC, Dash AP (2009) Identification of malaria hot spots for focused intervention in tribal state of India: a GIS based approach. Int J Health Geogr 8:30

Taylor PJ (1977) Quantitative methods in geography. Houghton Miffin, Boston

Teodoro PE, de Oliveira-Júnior JF, da Cunha ER et al (2016) Cluster analysis applied to the spatial and temporal variability of monthly rainfall in Mato Grosso do Sul State, Brazil. Meteorol Atmos Phys 128:197

Tobler WR (1970) A computer movie simulating urban growth in the Detroit region. Econ Geog Suppl 46:234–240

Tomlin CD (1990) Geographic information systems and cartographic modelling. Prentice-Hall, Englewood Cliffs, New Jersey

Waller LA, Lawson AB (1995) The power of focused tests to detect disease clustering. Stat Med 14:2291–2308

WHO (2010). http://www.who.int/malaria-/world_malaria_report_-2010/worldmalariareport2010.pdf. Accessed 15 Feb 2012

Wilson AG, Bennett RJ (1985) Mathematical methods in human geography and planning. Wiley, London

Worboys MF (1992) A generic model for planner geographical objects. Int J Geogr Inf Syst 6:353–372

World Bank (2007) Jharkhand—addressing the challenges of inclusive development. Poverty reduction and economic management India country management Unit South Asia. Report No. 36437-IN

Yuan M (1996) Temporal GIS and spatio-temporal modeling, published on CD-ROM by the national center for geographical information and analysis

Chapter 4
Spatial Statistics and Public Health Events

4.1 Introduction

Statistical investigation which covenants with spatial or spatio-temporal datasets is called as the science of spatial statistics or geostatistics. The term geostatistics was first proposed by Hart (1954) in a geographical framework, in relation to acmes the application of specific statistical methods to remark covering a regional distribution. According to G Matheron (Father of geostatistics), geostatistics is "*the application of probabilistic methods to regionalized variables which designates any function displayed in a real space*". Oleas (1991, p. 31) defines that geostatistics as "*the application of statistical methods.... for use in the earth sciences, particularly in geology*". Deutsch and Journel (1992) defines that "*geostatistics is the study of phenomena that varies in space and/or time*". Caers (2005) stated that "*geostatistics can be defined as the branch of statistical sciences that studies spatial/temporal phenomenon and capitalizes on spatial relationships to model possible values of variable(s) at unobserved, unsampled locations*". Geostatistics is employed for exhibiting spatial data. It offers precise and consistent assessments of phenomenon at sites where no measurements are accessible.

Spatial statistical study was first established in the 1950s as an outcome of interest in a real or block averages for ore reserves in the mining industry. It is an invaluable tool invented from the mining and petroleum industries, beginning with the work by Danie Krige in the 1950s and was further established by Georges Matheron in the 1960s. Geostatistics has its beginnings as a technical field in the work of Youden (1951), Matheron (1963) and Gandin (1963). It was well along-proven as a scientific field by Journel and Huijbregts (1978), Journel (1986), Isaaks and Srivastav (1989), Cressie (1993), Stein (1999) and many other authors. Geostatistical studies are used in the fields where the data are observed as point observation and forecast (at unsampled sites) is anticipated. Recently, geostatistical studies are conducted in various branches of geography, particularly with hydrological data (Skøien et al. 2006; Holliger et al. 2008; Dionissios 2015), in mining

G. S. Bhunia and P. K. Shit, *Geospatial Analysis of Public Health*,
https://doi.org/10.1007/978-3-030-01680-7_4

applications (Zeqiri et al. 2012; Fouedjio 2016), the development of spatial networks (Bhat et al. 2015; Rushworth et al. 2015), air quality studies (Enkhtur 2013), soil science data (Bhunia et al. 2016; Tola et al. 2016), biological applications (Relethford 2008), environmental management (Shit et al. 2016; Wu et al. 2011), and public health, particularly those in the spread of diseases (Bhunia et al. 2013).

4.2 Use of Spatial Statistical Methods in Public Health Events

Spatial epidemiology is an integration of a diversity of fields attaining knowledge from epidemiology, statistics, public health, geography as the study distribution. Coming to public health, the spatial statistics techniques provide imperative information on how a disease is extended; which are the regions affected by the disease and forecast the next regions which have higher prospect to be affected in order to control it (Elliott et al. 2000). The investigation of public health data employs spatial statistical methods to determine clusters or differences coming from different geographic locations (Shyti and Fetahu 2015). However, the prime role of probabilities stimulates the use of statistical methods to examine public health data and the use of spatial statistical approaches to appraise differences in rates observed from different geographic areas; disconnect pattern from noise, recognize disease clusters, and appraise the connotation of potential exposures (Bailey 2001). Several issues of major statistical journals have been devoted to spatial statistical methods in health applications (*American Journal of Epidemiology, BMC infectious Disease, Parasite and Health, Journal of Royal Statistical Society, Geospatial health, Memorias do Instituto Oswaldo Cruz, Geographical Analysis, IEEE Transactions on Geoscience and Remote Sensing, Statistics in Medicine and Statistical Science*). However, the epidemiological and public health implications and benefits arising from the use of such methods are more difficult to assess.

4.2.1 Mapping Disease Rate

The disease rate maps are created from assessments of incidence or prevalence, while importance maps are based on the practice of statistical distribution. There are various ways in which data can be represented on maps (Few 2009). These comprise dot-density maps, proportional circles/spheres, choropleth maps, contour maps etc. (Sarah et al. 2016). The analysis of the spatio-temporal distribution of the incidence of disease and its association to possible risk factors has a significant part to play in numerous types of public health and epidemiological investigations. In geographical epidemiology, there are four broad areas of statistical interest are acknowledged:

- *Disease mapping*: Disease mapping emphasizes on creating a map of the true underlying spatial distribution of the disease incidence, assumed noisy detected data on incidence rates. This may be beneficial in signifying premise for further examination or as part of general health observation and the observing of health complications.
- *Ecological studies*: Ecological studies are apprehensive with reviewing relations between observed incidence of disease and possible risk factors as sedate on clusters. Such investigations are appreciated in inspecting the etiology of infection and may aid to mark supplementary research and possibly deterrent measures.
- *Disease clustering studies*: This study concentrates on ascertaining ecological areas with substantial increasing risk of disease or on evaluating the signal of high risk around assumed sources of hazard.
- *Environmental assessment and monitoring*: This is concerned with determining the spatial distribution of environmental factors pertinent to health and acquaintance to these so as inaugurate essential controls or yield deterrent activities.

4.2.2 *Measuring the Geographical Distributions of Disease Pattern*

Meanwhile different procedures maneuver at different scales and over altered areas in ecology and epidemiology, problems of scale and extent are essential to geo-statistical analysis. Exploratory analysis of spatial data sets may support detect patterns at various scales. Determining the distribution of a set of features allows analyzing a value that denotes a center, compactness and/or orientation of the distribution (Table 4.1).

4.2.2.1 Mean Center

Mean is an imperative measure of central tendency for a set of data which is measured from the average x and y co-ordinates of all input features within the area of interest (Mitchell 2005). It is beneficial for pursuing vicissitudes in the distribution or for associating the disseminations of dissimilar forms of features. The mean center can be calculated by separately averaging the X and Y co-ordinates as follows:

$$\overline{\mathrm{X}}_c = \frac{\sum X_i}{n} \text{ and } \overline{\mathrm{Y}}_c = \frac{\sum Y_i}{n}$$

Table 4.1 List of some important geostatistical software package

Software	Application	Source code
GsLib	A classical open-source package dedicated to geostatstics	FORTAN 77 and 90
SGeMS	An open source package dedicated to geostatisrics with user-friendly interface	C++
Isatis	A complete proprietary solution for geostatistics and resource estimation	
Geostat	A complete proprietary solution for geostatistics, geological modeling and resource estimation	
D-STEM	Able to handle spatio-temporal univariate and multivariate dataset	MATLAB
GEOPACK	Allows both novice and advanced users to undertake geostatistical analyses (linear regression, polynomial regression, and Kolomogorov-Smirnov tests; inner estimations and nonlinear estimations) of spatially correlated data	MS FORTAN

where,

$\overline{\mathrm{X}}_c$ mean center of X; $\overline{\mathrm{Y}}_c$mean center of Y;
X_i X co-ordinate of point i, Y_i = Y co-ordinate of point i;
N number of features in the distribution.

In location co-ordinate system, it is appropriate to assign differential weights to points in a geographical location that is analogous to frequencies in the analysis of grouped data.

$$\overline{\mathrm{X}}_{wc} = \frac{\sum f_i X_i}{f_i} \text{ and } \overline{\mathrm{Y}}_{wc} \frac{\sum f_i Y_i}{f_i}$$

where,

$\overline{\mathrm{X}}_{wc}$ weighted mean center of X; $\overline{\mathrm{Y}}_{wc}$weighted mean center of Y.
f_i frequency of point i.

Potential applications:

- It is considered as the center of gravity of a point pattern or spatial distribution.
- Appropriate to assign differential weights to points in a spatial distribution.
- Serves as a spatial analogue to the mean that minimizes the sum of squared deviations of a set of points.

4.2.2.2 Median Center

To determine the central geographic location that minimizes the sum of unsquared distances from all other points in a spatial distribution to that central location, median center is used. The co-ordinates of the weighted median center will minimize the expression:

$$Median = \sum f_i \sqrt{(X_i - X_{we})^2 + (Y_i - Y_{we})^2}$$

where,

X_{we} and Y_{we}	Weighted Euclidean median.
F	represents the population or any other features appropriate o the spatial problem

Potential applications:

- Minimizes the overall Euclidean distance to the features in a dataset.

4.2.2.3 Standard Distance

Standard distance measures the extent of absolute dispersion in a geographical distribution. The standard distance statistics integrates the straight-line or Euclidean distance of each point from the mean center and can be represented on a map by drawing a circle with the radius equal to the standard distance value. The standard distance tools build a circle polygon. The standard distance can be calculated as:

$$S_D = \sqrt{\frac{\sum(X_i - \overline{X}_c)^2 + \sum(Y_i - \overline{Y}_c)^2}{n}}$$

where, X_i and Y_i are the co-ordinates for features i; $\{\overline{x}_c, \overline{y}_c\}$ represents the mean center for the features, and n is the equal to the total number of features.

Potential applications:

- Comparing the values of two or more distribution and to develop strategies for addressing the events.
- Likening the same type of features over different time periods.
- Relating the distributions of features to stationary features.

4.2.2.4 Standard Deviation of Ellipse (SDE)

The method calculates the standard deviation of the x-coordinates and y-coordinates from the mean center to define the axes of the ellipse. The ellipse aids to determine

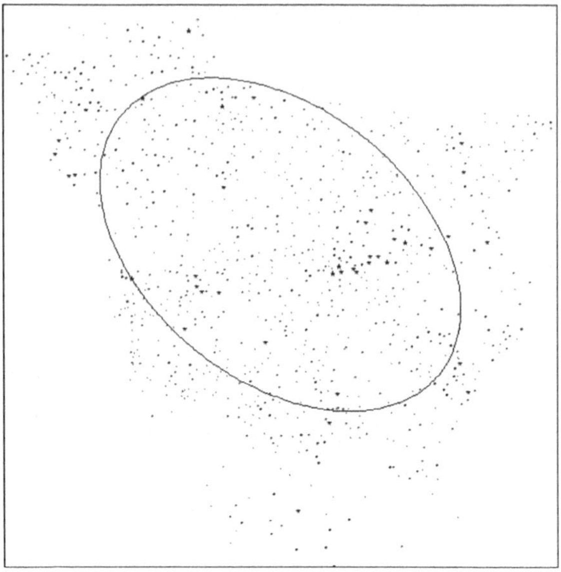

Fig. 4.1 Schematic diagram of standard deviation of ellipse of visceral leishmaniasis distribution in Vaishali district

the distribution of features is elongated and hence has a particular orientation (Mitchell 2005). The SDE is used the geographic locations of the features or the locations influenced by an attribute value associated with the features (Fig. 4.1). The SDE for x-axis and y-axis can be calculated as:

$$\sigma_x = \sqrt{\frac{\sum_{i=1}^{n} (\bar{x}_i cos\theta - \bar{y}_i sin\theta)^2}{n}}$$

$$\sigma_y = \sqrt{\frac{\sum_{i=1}^{n} (\bar{x}_i sin\theta - \bar{y}_i \cos\theta)^2}{n}}$$

where, x_i and y_i are the co-ordinates for feature I, $\{\bar{x}, \bar{y}\}$ represents the mean center for the features, and n is the equal to the total number of features.

Potential applications:

- Mapping the distributional trend of the set of features aids to identify the relationship to particular physical features.
- Comparing the size, shape and overlap of ellipses for various groups to determine the segregation among the groups.
- Plotting ellipses for a phenomenon over time may be used to model its spread.

4.2.3 Spatial Clustering Methods

Spatial clustering is a vital constituent of geographical data mining refers to the groups of similar objects into classes (Han and Kamber 2006). In spatial data sets, clustering permits a generalization of the spatial component like explicit location and extension of spatial objects which define implicit relation of spatial neighborhood (Bindiya et al. 2013). It is commonly used in disease surveillance (Goovaerts 2006), spatial epidemiology (Auchincloss et al. 2012), population genetics (Diaz-Lacava et al. 2015), landscape ecology (Jensen et al. 2006), crime analysis (Kerry et al. 2010) and many other fields. The clustered analysis may be event-based, population based, field-based or feature-based. Event-based data comprise point locations (i.e., period of diagnosis of cases or disease in people) and counts (i.e., occurrence of disease at particular location). Population-based data integrate information on the population from which the events ascended, and embrace disease rates with case counts in the numerator and size of at-risk population in the denominator. Field-based data are observations that are incessantly disseminated over space. Feature-based data take account of administrative boundaries and representative spatial features (point, line and polygon) that may be derived from the field-based data.

Cluster analysis acts key roles in the edifice of spatial models and in exploratory spatial data analysis (ESDA). For feature based data, a cluster might be a spatial aggregation of boundaries. Model development requires an appreciative of the patterns of spatial discrepancy as one often desires to integrate germane features of attribute variability into the model. Spatial clusters are of interest because they are the trace of space-time processes that are focus of geographic studies.

Statistical pattern recognition ensues by calculating a cluster, outliers, autocorrelation, hotspot, cold spot, trend and boundaries that quantify the relevant aspect of spatial pattern in event-based, population-based, field-based or feature-based. Waller and Jacquez (1995) solemn this method by recognizing five constituents of a test for a spatial pattern:

- The *test statistics* enumerates a pertinent aspect of spatial pattern, i.e., Moran's I, Geary's c, LISA etc.
- The *alternate hypothesis* defines the spatial pattern that the rest is intended to perceive. This may be a precise alternative, such as clustering close to focus.
- The *null hypothesis* designates the spatial pattern predictable when the alternative hypothesis is deceitful i.e., uniform pattern of distribution.
- The *null spatial model* is an apparatus for producing the reference distribution i.e., based on the distribution theory or may use randomization techniques (Monte Carlo).
- The *reference distribution* is the dissemination of the test statistics when the null hypothesis is true.

4.2.3.1 Join Counting

The representation and analysis of the spatial association of groups (e.g., nominal, qualitative, categorical, binary data) can be attained using *join count statistics* (Moran 1948). This technique was first proposed by human geographers to check whether or not contiguous countries presented a spatial pattern of disease. Geary (1954) developed join count statistics based on the probability that neighboring spatial units were of the same type. A '*join*' is literally a connection between two points that can be joined via a connection matrix. Join counting is a measure of clustering appropriate to nominal area data where the actual and expected number of joins between areas with dissimilar values is estimated (Hay et al. 2002). The smaller the ratio of dissimilar to total joins the more clustered the disease pattern. Cliff and Haggett (1988) applied join count statistics to establish a very high degree of cholera cases in London.

Potential application:

- It is not necessary to convert the area data to point representations, but does not lend itself to elucidating the size or pattern of clustering.

4.2.3.2 Geographical Analysis Machine

The Geographical Analysis Machine (GAM), an explanatory method was first introduced by Openshaw and others in 1987 to reconnoiter datasets for substantiation of spatial patterns, (Openshaw et al. 1987). The technique was established at the Department of Geography at the University of Newcastle in the mid-1980s in an endeavor to govern whether spatial clustering was apparent in the incidence of various cancers in children in northern England. GAM practices a sequence of overlapping circles in which observed an expected numbers of cases are calculated for circles, of a range of radii with centers at every point of a fine grid.

4.2.3.3 Nearest Neighbor Analysis

Nearest neighbor analysis (ANN) studies the spatial association and distribution between each point and point adjoining to it. It is technique of reconnoitering patterns in locational data by associating realistically the observed circulation functions or event-to-event or random point-to-event nearest neighbor distances. It was formerly introduced by P. Hertz but its systematic use was initiated by Clark and Evans, the two American Plant Ecologist. The nearest neighbor ratio (NNR) is calculated as the observed average distance divided by the expected average distance which is mathematically expressed as:

$$ANN = \frac{\overline{D}_O}{\overline{D}_E}$$

where $\overline{D}_0$ is the observed mean distance between each feature and its nearest neighbor, can be calculated as:

$$\bar{D}_o = \frac{\sum_{i=1}^{n} d_i}{n}$$

where d_i refers to equal distance between feature i and its nearest neighboring feature.

N corresponds to the total number of features.

$\overline{D}_E$ is the expected mean n distance for the features given in a random pattern, can be calculated as:

$$\overline{D}_E = \frac{0.5}{\sqrt{\frac{n}{A}}}$$

where A is the area of a minimum enclosing rectangle around all features, or user specified area value.

The ratio of R varies from 0 (completely clustered) to 1 (random) to 2.149 (completely dispersed). In this analysis, the distribution was considered dispersed if the mean of the observed nearest neighbor distance was found to be greater than that of a random pattern ($R > 1$), while clustering was indicated when the mean of the observed nearest neighbor distance was found to be less that of a random pattern ($R < 1$).

The average nearest-neighbor z-score for the statistics is calculated as:

$$Z = \frac{\overline{D}_{obs} - \overline{D}_{ran}}{SE_r}$$

where is the expected mean nearest neighbor distance in a random pattern and the standard error of the mean nearest neighbor distance i.e.,

$$SE_r = \frac{0.26136}{\sqrt{\frac{n^2}{A}}}$$

If $Z_r > 1.96$ or $Z_r < -1.96$ (probability < 95%), we concluded that the calculated difference between the observed pattern and the random pattern was statistically significant. Alternatively if $-1.96 < Z_r < 1.96$ (probability < 95%), we concluded that the observed pattern was not significantly different from a random pattern even if the point appeared visually clustered or dispersed (Fig. 4.2).

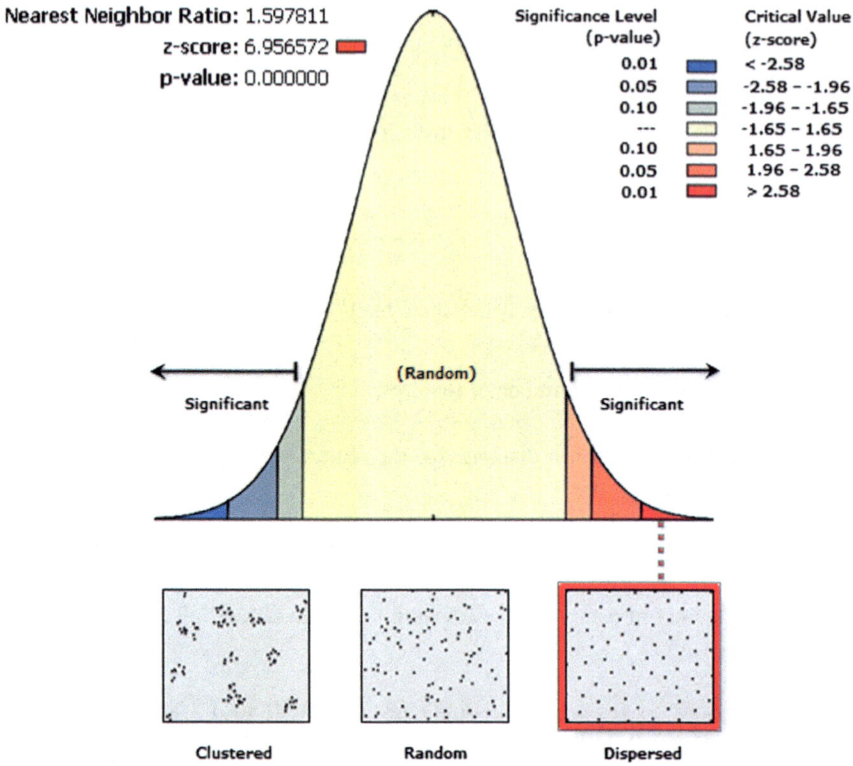

Fig. 4.2 Schematic diagram of average nearest distance summary analyzed through ArcGIS software v10.0

4.2.3.4 Spatial Autocorrelation

Spatial autocorrelation denotes to the interdependence of values of a variable at various geographic locations and measure the nature and strength of that interdependence (Getis 2008). Tobler's first law of geography summarizes this circumstance: "*everything is related to everything else, but near things are more related than distant things*" (Tobler 1970). Conversely, spatial autocorrelation refers a colony subsists between values of a variable in adjoining or remote places, or a systematic pattern in values of a variable across the locations on a map because of essential collective factors. These adjoining values can be recognized by an n-by-n binary geographic connectivity/weights matrix. Hubert et al. (1981) provides a concise definition:

> Spatial autocorrelation refers to the relationship between some variable observed in each of the n localities and a measure of geographical proximity defined for all n(n − 1) pairs chosen from n (pp. 224)

Until 1968, spatial autocorrelation had been entitled 'spatial dependence, 'spatial association', 'spatial interaction', and 'spatial dependence'. Moran (1950) had used the word spatial correlation and Whittle (1954) denotes comprehensively to autocorrelation in a spatial milieu. The 1968 conference paper was published in 1969 under the title "*The Problem of Spatial Autocorrelation*" (Cliff and Ord 1969). The monograph *Spatial Autocorrelation* by Cliff and Ord was published in 1973. The concept of spatial autocorrelation has been long in coming resembles with the ever-growing number and the kind of models that have been used to explore phenomena in numerous academic fields (Griffith 2009). Conceivably the most significant contribution was to elucidate and simplify Moran's prior work. The instants of Moran's distribution, called *Moran's I* were entirely established by Cliff and Ord under varying sampling norms.

Spatial autocorrelation may be categorized as either positive or negative. In a positive case, geographically adjacent values of a variable tend to be similar on a map, e.g., high values tend to be located near high values, medium values near medium values and low values near low values. Most social science variables tend to be moderately positively spatially auto-correlated because of the way phenomenon is geographically organized. While a negative spatial autocorrelation has different values seeming in adjacent association and zero spatial autocorrelation means geographically random phenomenon and chaotic landscapes.

Therefore there are two primary reason s to measure spatial autocorrelation—(i) it indexes the nature and degree to which an important statistical hypothesis is desecrated and also designates the amount to which conventional statistical implications are bargained when non-zero spatial autocorrelation is ignored; (ii) the extent of spatial autocorrelation defines the overall pattern across a geographic landscape, assistant spatial prophecy and permitting recognition of striking aberrations.

4.2.3.5 Global Cluster Statistics

Global cluster statistics are subtle to spatial clustering or exoduses from the null hypothesis that happens any place within the study area. A popular measure of spatial autocorrelation is Moran's I (Moran 1950; Griffith 1987).The Global Moran's I is an inferential statistics that aids to define the degree of resemblance between each areal unit and it adjoining neighbors. It evaluates whether the pattern expressed is clustered, dispersed or random (Mitchell 2005). The Moran's I statistics can be computed as:

$$I = \frac{\sum_{i=1}^{n} \sum_{j=1}^{m} w_{ij}(x_i - \overline{x})(x_j - \overline{x})}{s^2 \sum_{i=1}^{n} \sum_{j=1}^{m} w_{ij}}$$

where x_i is the value at a point i, x_j is the value at point i's neighbor j, w_{ij} is a coefficient, n is the number of pints, and s^2 is the variance of x values with a mean of $\bar{x}$. The co-efficient w_{ij} is the weight for measuring spatial autocorrelation. Typically w_{ij} is defined as the inverse of the distance (*d*) between points *i* and *j* or $1/d_{ij}$.

The value of *Moran's I* fall between −1.0 and +1.0. If the values in the dataset incline to cluster spatially (high values cluster close other high values, low values cluster near other low values), the *Moran's I* will be positive. When high values deter other high values, and be likely to near low values, the index will be negative. If positive cross-product values balance negative cross-product values, the index will be near zero.

The z-score directs the likelihood that the point pattern could be consequence of random chance. Z-scores are basically standard deviation. The z-score for the statistics is calculated as:

$$z_I = \frac{I - E[I]}{\sqrt{V[I]}}$$

where,

$$E[I] = \frac{-1}{(n-1)}$$

$$V[I] = E\left[I^2\right] - E[I]^2$$

The $E[I]$ approaches '0' when number of points *n* is large. *Moran's I* is close to $E[I]$ if the pattern is random. It is greater than $E[I]$ if contiguous points lean towards similar values (i.e., spatial correlated) and less than $E[I]$ if neighboring points lean towards different values (i.e., not spatially correlated).

Consequently, p-value determines the probability. For the pattern analysis tool, observed spatial pattern was created by some random process. When the p-value is very small, it means it is very unlikely that the observed spatial pattern is the result of a random process, so you can reject the null hypothesis (Fig. 4.3).

Very high or very low (negative z-score), associated with very small p-values, and are found in the tails of the normal distribution. When we run a feature pattern analysis tool and it yields small p-values and either a very high or a very low z-score, this indicates it is unlikely that the observed spatial pattern reflects the theoretical random pattern represented by the null hypothesis (Ebdon 1985; Mitchell 2005).

For example, figure portrays the visceral leishmaniasis distribution the point distribution produces a Moran's I of 0.369951, which is much higher that $E[I]$ of −0.027778, and a z-score of 4.0737763, which suggest that the likelihood of the pattern being a result of random chance (Fig. 4.4).

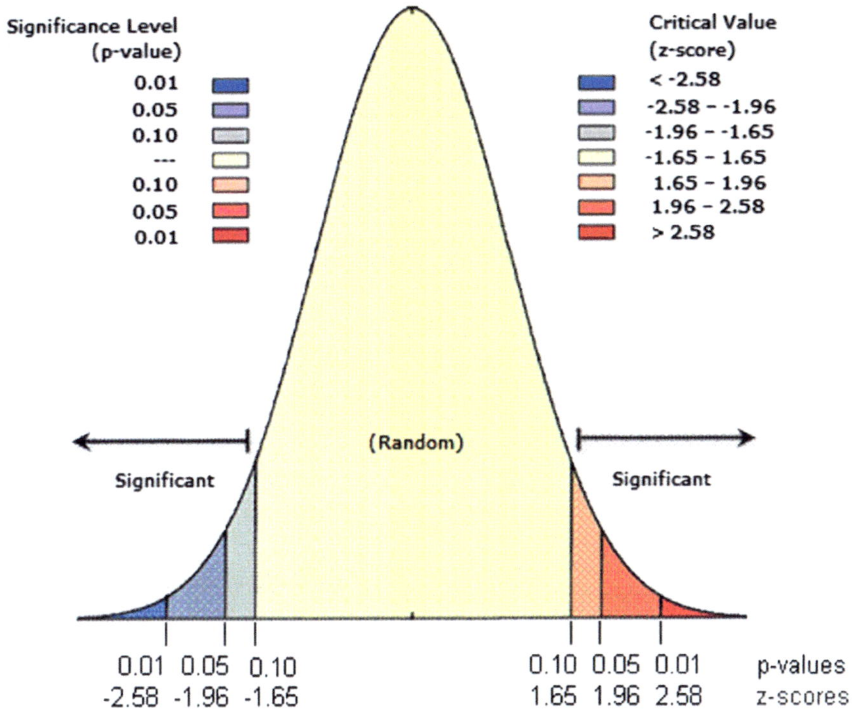

Fig. 4.3 Schematic diagram associated with the Z-score and p-value (*Source* Mitchell 2005; ArcGIS Software v10.0, ESRI, Redlands, California, USA)

Potential application:

- Help to ascertain a suitable neighborhood distance for a variation of spatial analysis method where spatial autocorrelation is stoutest.
- Abridge the diffusion of an idea, disease or trend over space and time enduring quarantined or concerted or scattering and becoming more diffuse.
- Measure extensive tendency of segregation over time.

4.2.3.6 Incremental Spatial Autocorrelation

Incremental spatial Autocorrelation (ISA) uses Moran's I to test for spatial autocorrelation within distance bands. This measures intensity of spatial clustering for a succession of distance and optionally generates a line graph of those distances. The intensity of clustering is determined by the z-score. Peak in the line graph imitate the distances where the spatial procedures endorsing clustering are most prominent (Maingi et al. 2012). Distance of first peak in z-score is usually used for further analysis and also useful for determining the appropriate scale for further analysis.

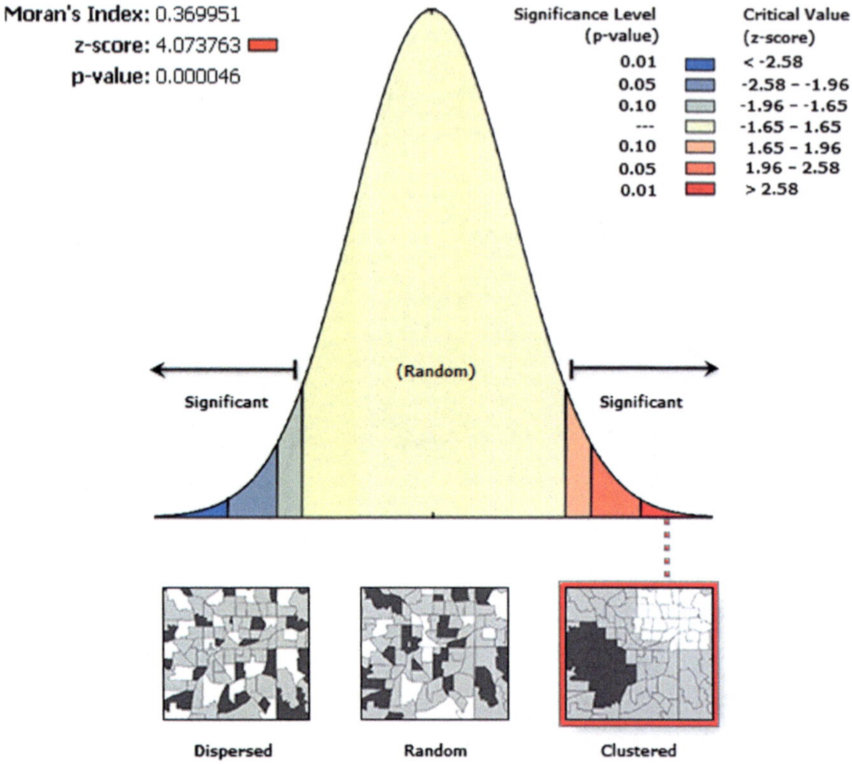

Fig. 4.4 Schematic diagram of spatial autocorrelation report of visceral leishmaniasis distribution in Bihar analyzed through ArcGIS software v10.0 (ESRI, Redlands, California, USA)

ISA also determine a subsample should be taken to remove autocorrelation. de la Cruz et al. (2014) uses ISA to evaluate bovine tuberculosis in Madrid, Spain (2010–2012).

4.2.3.7 Cluster and Outlier Analysis

Local Indicators of Spatial Autocorrelation (LISA) quantify the spatial auto-correlation and clustering within the small areas that together comprise the study area (Anselin 1995). The Local Moran's I statistics of spatial association is computed as:

$$L_i = Z_i \sum W_{ij} Z_j$$

Here, L_i is the LISA statistics for area i, z is the observation at location i, scales to have a mean of 0 and the unit standard deviation (z-score); W_{ij} is the spatial weight between feature i and j. W represents the spatial association between site i and j; Z represents the association of values of a random variable at site i and j.

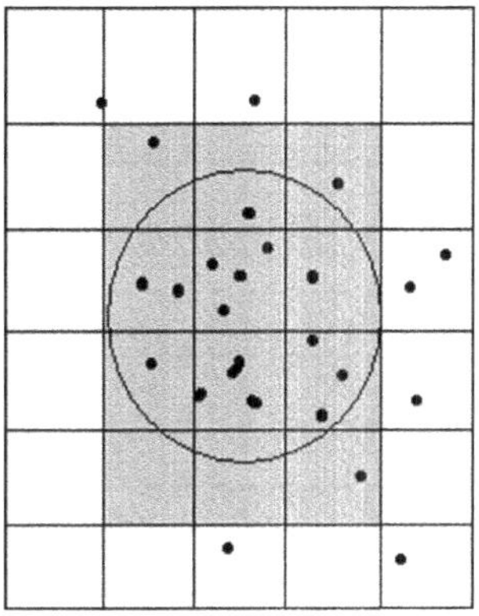

Fig. 4.5 Units within a specified radius

A positive value of I direct that a feature has adjoining features with correspondingly high or low attribute value, this feature is a part of cluster. A negative value of I specifies that a feature has adjoining features with different values; this feature is an outlier (Fig. 4.5). The outcome of LISA differentiates between a statistically significant ($P < 0.05$) cluster of high values, cluster of low values, outlier in which a high value is bounded mainly by low values, and outlier in which a low value is encircled primarily by high values.

Potential application:

- Tools help to identify concentration of high values or low values and spatial outliers. This analysis also aids to determine the anomalous spreading pattern and/or sharpest boundaries between affluence.

4.2.3.8 Focused Statistics

Focused statistics enumerate the clustering around a definite locality called focus. This test is predominantly beneficial for reconnoitering possible clusters of disease adjacent latent sources of environmental pollutants. Lawson (1989) planned this test that score each area for the change between observed and expected disease counts, can be calculated as:

$$U = \sum_{i=1}^{N} g_i(o_i - e_i)$$

Here, N refers to area, g is a function of important the exposure to the focus, o_i is the observed number of features in area i, and e_i is expected number if features in that area. A commonly used exposure function is the inverse distance to the focus (1/d). The null hypothesis is no clustering relative to the focus and the expected feature count. Thus this is intended as the poisson anticipation by the population at risk in each area and the hypothesis that disease risk is uniform pattern.

4.2.3.9 Hot-Spot Analysis

Hotspot analysis is a spatial cluster recognition technique which ascertains statistically significant spatial concentration high values or low values concomitant with a set of geographic features (Getis and Ord 1992). To measure the hotspot analysis *Getis-Ord* G_i^* statistics is used. The hotspot analysis tool performs by observing at each feature within the framework of adjacent features. If the value for a feature is high, and the values for adjoining features are also high, hotspot may be exhibited. The basic statistics is defined as:

$$G_i(d) = \frac{\sum_j w_{ij}(d)x_j}{\sum_j x_j}$$

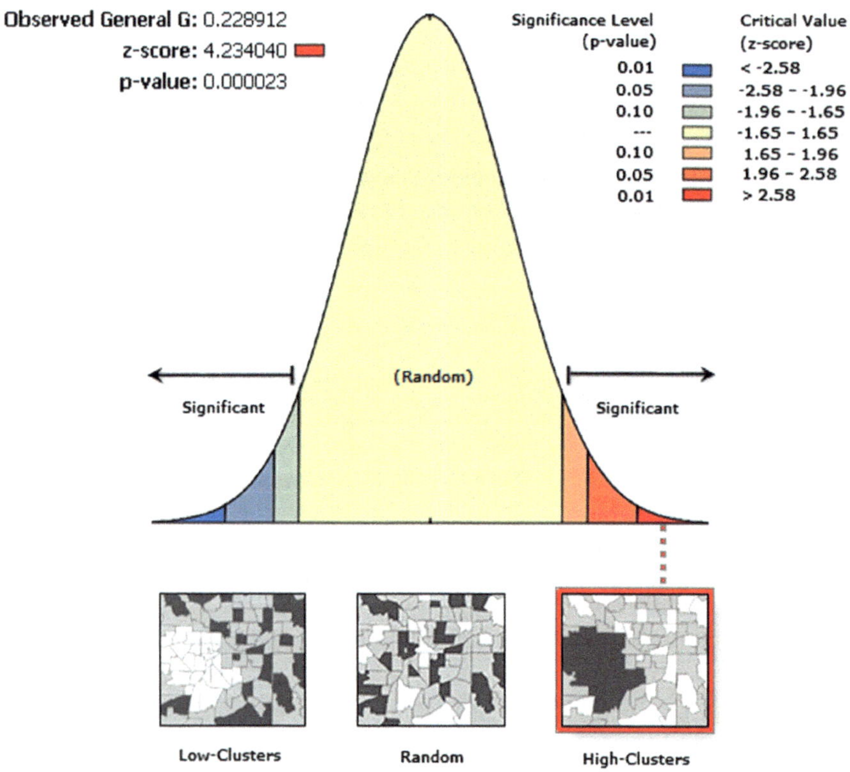

Fig. 4.6 Schematic diagram of Getis-ord Gi* statistics of visceral leishmaniasis distribution in Bihar (India)

In this equation, the x_j is the weighted values or the points within the study area, w_{ij} is the binary symmetric weights matrix with ones for all points j with in the distance d of point i and 0 otherwise.

In Getis-Ord statistics, G_i^* is asymptotically usually distributed as d increases. Where, n is very large, a conservative approach would be to choose d so that the number of neighbors is at least. When n is comparatively small, as few as its neighbors could be used without severe inferential error unless the essential distribution is much skewed (Fig. 4.6).

Potential application:

There are several potential application in crime analysis, epidemiology, voting pattern analysis, economic geography, retail analysis, traffic incident analysis and demographies. Hotspot analysis helps to determine:

- Summarizing the level at which spatial phenomenon cluster o scrutinize variations at different items or in different locations.
- Likening spatial pattern of dissimilar types repel competition.
- Look for unpredicted spikes in the number of emergency which might signpost an occurrence of a local or regional problem.

4.2.4 Conceptualization of Spatial Relationship

A number of spatial relationship model are available with the clustering analysis in GIS software. These are as follows

- *Inverse distance method*: This is most suitable with continuous data or to model developments where the nearer two features are in space, the more they intermingle/affect each other.
- *Fixed distance method*: The fixed distance method is best suited for polygon data where there is a large dissimilarity in polygon size.
- *Zone of indifference*: The zone of indifference conceptualization performs well when fixed distance applicable, but magnificent shrill boundaries on neighborhood associations are not a perfect depiction of the data.
- *Polygon contiguity*: This is effective when polygons are analogous in size distribution and when spatial associations are a function of polygon proximity.
- *K-nearest neighbor*: This approach is effective when a minimum number is neighbor available for the analysis. When the fixing the scale of analysis is less significant than fixing the number of neighbors, the K-nearest neighbor method is appropriate.
- *Delaunary triangulation*: The method is appropriate when the data includes island polygon (isolated polygon do not share boundaries with other polygon) or if there is very uneven spatial distribution of features.

4.2.5 *Spatial Statistical Model*

Spatial statistics is a set of models and tools developed for statistical analysis of continuous data. The main objective of applying spatial statistical model to disease data is finally to succor cognizant decision making for disease intervention. Once the dissemination of disease has been drawn and examined the patterns of that distribution, attempted to explicate the casual factors to choose how can variation these patterns to improve the health of people. These data can be measured at any location in space, but they are available in a limited number of sampled points. This section starts with some geostatistical model function in association with public health application.

4.2.5.1 GIS Based Interpolation Methods

Of the six GIS based interpolation methods considered e.g., *Inverse distance weighted* (IDW), *Radial basis function (*RBF), *Global polynomial interpolation* (GPI) and *Local polynomial interpolation* (LPI) are referred to as the deterministic interpolation methods, while kriging and co-kriging are referred as the geostatistical interpolation methods. The deterministic interpolation methods generate a surface from the measured data points based on the degree of similarity and smoothing, while the geostatistical methods generate a surface using a statistical properties of the measured data points after consideration of spatial configuration of the measured data points around the prediction location (Kumar et al. 2014). Based on the data points used in generating a surface, one can further classify the deterministic interpolation method used all the measured sample data points in making the predictions, while the local interpolation method uses only a smaller sample of the measured data points from the neighbourhood within a larger area. The GPI is categorized in the global interpolation method, while the IDW, RBF and LPI are categorized as the local interpolation methods. Alternatively, one can also categorize the deterministic methods based on whether the surface generated passes through the measured points, the interpolation methods are referred as the inexact interpolators. The IDW and the RBF are the exact interpolators, while the GPI and the LPI are the inexact interpolators.

Deterministic Methods

Deterministic approaches governed by the degree of resemblance of the values or the extent of smoothing in the surface. These methods are not depending upon a random spatial process model, and there is no categorical dimension or modeling of spatial auto-correlation in the data.

- *Inverse distance weighted (IDW) Interpolation*

IDW interpolation explicitly depends on the first law of Geography. To envisage a value for some unmeasured location, IDW will employ the estimated values adjacent the estimate location. Measured values that are nearby to the prediction location will have the greater impact (i.e., weight) on the projected value at that unidentified point than those that are further away (Fisher et al. 1987). Thus IDW undertakes that each estimated point has a local impact that reduces with distance (or distance to the power of $k > 1$), and weights the points nearer to the prediction location greater than those farther away. The power k controls the degree of local influence. A power of 1.0 means a constant rate of change in value between points (linear interpolation). A power of 2.0 or higher suggests that the rate of change in values is higher near known points and levels off away from it. The degree of local influence also depends on the number of known points used in the estimation. The general equation for the IDW method is:

$$Z_0 = \frac{\sum_{i=1}^{s} z_i \frac{1}{d_i^k}}{\sum_{i=1}^{s} \frac{1}{d_i^k}}$$

where z_0 is the estimated value at point 0, z_i is the z value at point i, d_i is the distance between point i and 0, s is the number of known points used in estimation, and k is the specified power. An important characteristic of IDW interpolation is that all predicted values are within the range of maximum and minimum values of the known points. The advantage of IDW is that it is intuitive and efficient and it works best with evenly distributed points and it is sensitive to outliers (Balakrishnan et al. 2011).

- *Global polynomial interpolation (GPI)*

The interpolation fits a smooth surface that is defined by a mathematical function to the input sample points. It is best used for surfaces that change slowly and gradually. As such the GPI is useful for creating smooth surfaces and identifying long-range trends in the dataset. A 1st order GPI fits a single plane through the data; a 2nd order GPI fits a surface with a bend in it, permitting the calculation of surfaces demonstrating valleys; a 3rd GPI permits for 2 bends; and so forth. Conversely, when a surface has a dissimilar form, as in a landscape that slopes, levels out, and then slopes again, a single GPI will not fit well (Johnston et al. 2001)

- *Local polynomial interpolation (LPI)*

LPI fits the local polynomial by points only within the definite neighborhood instead of all the data (Hani and Abari 2011). Then the neighborhoods can overlay, and the surface value at the center of the neighborhood is assessed as the prophesied value. LPI is proficient of generating surfaces that seizure the little range disparity (ESRI 2001).

LPI fits the definite order (zero, first, second, third and so on) polynomial using points only within the defined region. The neighborhoods overlap, and the values used for each prediction is the value of the fitted polynomial at the center of the neighborhood.

- *Radial basis function (RBF)*

RBF envisages values matching with those measured at the same point and the produced surface entails transient through each estimated points. The predicted values can differ above the maximum or below the minimum of the measured values (Bhunia et al. 2016; Li et al. 2011). RBF appropriate a surface through the estimated sample values while reducing the total arch of the surface (Johnston et al. 2001). RBF technique is a family of five deterministic precise interpolation methods: thin-plate spline (TPS), spline with tension (SPT), completely regularized spline (CRS), multi-quadratic function (MQ and inverse multi-quadratic function (IMQ). RBF is unsuccessful when there is an intense alteration in the surface values within small spaces (ESRI 2001).

Geostatistical Methods

Geostatistical methods assume that at least certain spatial difference observed in natural phenomena can be modeled by random processes with spatial autocorrelation and need that the spatial autocorrelation be openly displayed.

- *Krigging*

Krigging is a linear interpolation technique that offers linear equitable approximation for quantities, which differ in space and it is an advanced geostatistical method that creates and assessed surface from a distributed set of points varies from a dispersed set of points with z-values (Gunarathna et al. 2016). Krigging undertakes that the spatial discrepancy of an attribute such as variations in grade within an ore body is neither entirely random nor deterministic. Instead, the geographical dissimilarity may contain of three constituents: a spatially associated constituent, demonstrating the difference of the decentralized variable, and a random error term. The common equation for the *krigging* technique is:

$$Z(S_o) = \sum_{i=1}^{N} \lambda_i Z(S_i)$$

$Z(S_i)$ is the measured value at the ith location; λ_i is the unknown weight for the measured value at the ith location; S_o is the prediction location; N is the number of measured value.

- *Ordinary krigging (OK)*

Ordinary kriging technique includes statistical properties of the spatially autocorrelated component. Ordinary kriging assumes the model

$$Z(s) = \mu + \varepsilon(s)$$

where μ is an unknown constant. One of the main issues concerning OK is whether the assumption of a constant mean is reasonable. The *OK* method implements the semivariogram to express the spatial autocorrelation. The general equation for estimating the z value at a point is:

$$z_0 = \sum_{i=1}^{s} z_x W_x$$

where, z_0 is the estimated value, z_x is the known value at point x, W_x is the weight associated with point x, and s is the number of sample points used in estimation.

- *Universal Kriging (UK)*

UK assumes that the spatial variation in z values has a drift or a trend in addition to the spatial correlation between the sample points. UK assumes a trending mean and can be modeled as:

$$Z(s) = \mu(s) + \varepsilon(s)$$

where $\mu(s)$ is some deterministic function. UK incorporates the 1st order (plane surface) or 2nd order (quadratic surface) polynomial in kriging process. A 1st order polynomial is

$$M = b_1 x_1 + b_2 y_i$$

where M is the drift, x_i and y_i are the x-, y-coordinates of the sampled point i, and b_1 and b_2 are the drift co-efficient. A second order polynomial is:

$$M = b_1 x_1 + b_2 y_i + b_3 x_i^2 + b_4 x_i y_i + b_5 y_i^2$$

- *Co-kriging*

Cokriging is analogous to kriging, excluding that it uses additional covariates, e.g. the climatic variable and elevation. Conversely, *co-kriging* uses the compensations of inter-variable association. Co-kriging is most operative when the covariates are strongly correlated (Nalder and Wein 1998).

- *Semivariance*

Semivariance is a measure of the dissimilarity of the spatially correlated component. The range is the distance at which the spatial correlation vanishes, and the sill corresponds to the maximum variability in the absence of spatial dependence (Bhunia et al. 2016). The semivariance is the moment of inertia or extent of the scatterogram about the 45° (1–1) line revealed on the plot. Covariance and correlation are both processes of the resemblance of the head and tail values. The semivariance is computed by:

$$\gamma(h) = \frac{1}{2}\left[z(x_i) - z(x_j)\right]^2$$

where $\gamma(h)$ is the semivariance between known points, x_i and x_j, alienated by the distance h; and z is the attribute value.

The empirical semivariograms attained from the data is fitted by theoretical semivariogram models to yield geostatistical parameters, comprising nugget variance (c_o), structured variance (c_i), sill variance (c_o + c_i) and distance parameters (λ).

Sill: The semivariance value at which the variogram levels off. This is employed to denote the amplitude of a definite constituent of the semivariogram.

Range: The lag distance at which the semivariogram components reaches the sill value. Presumably, autocorrelation is essentially zero beyond the range.

Nugget: In theory the semivariogram value at the source should be zero. If it is considerably dissimilar from zero for lags very adjacent to zero, then this semivariogram value is stated to as the nugget. This nugget signifies inconsistency at distances smaller than the typical sample spacing, containing measurement error.

- *Empirical Bayesian kriging (EBK)*

EBK is a kriging-based interpolation technique that accounts for ambiguity in semivariogram approximation by pretending several semivariograms from the input data. Furthermore, EBKcan account for moderate non-stationarity by constructing local models on subsets of the input data (Pilz and Spöck 2007). For a given distance h, EBK uses a semivariogram model with the following form:

$$Power : \gamma(h) = Nugget + b|h|^{\alpha}$$

$$Linear : \gamma(h) = Nugget + b|h|$$

$$Thinplate\,spline : \gamma(h) = Nugget + b|h^2| * \ln(|h|)$$

The nugget and b (slope) must be positive, and α (power) must be between 0.25 and 1.75. This semivariogram model does not have a range or sill parameter because the function has no upper bound (Krivoruchko and Butler 2013).

Cross-Validation

Cross validation associates the interpolation procedures to appraise the model function. A cross validation study can benefit to choose between dissimilar weighting measures, between various search approaches, or between different estimation methods. The sample value at a particular location is temporarily discarded from the sample data set; the value at the same location is then estimated using the remaining samples. Once the estimate is calculated, the calculated value can be compared with the true value that was initially removed from the sample dataset. For determining the best interpolation method, three criteria of root mean square error (RMSE), mean error (ME) and %RMSE is used. The RMSE can be calculated as:

$$RMSE = \sqrt{\frac{1}{n}\sum\nolimits_{i=1}^{n}\left(Z_{i,act} - Z_{i,est}\right)^{2}}$$

where n is th number of points, $Z_{i,act}$ is the known value of point i, and $Z_{i,est}$ is the estimated value of point i.

ME is the indication of the degree of bias. The ME can be calculated as:

$$ME = \sum_{i=1}^{n}\frac{\left(Z_{i,act} - Z_{i,est}\right)}{n}$$

Positive value of ME show values more than real and negative values show estimates less than real. The more ME is smaller, the less the interpolation is skewed.

And %RMSE is the measures of sensitive to outliers. The %RMSE can be calculated as:

$$\%RMSE = \frac{RMSE}{\overline{x}} * 100$$

where, x is the mean of measured parameter.

4.3 Case Study: Spatial Statistical Analysis of Visceral Leishmaniasis (Kala-Azar) Incidence in Muzaffarpur District, Bihar (India)

4.3.1 Introduction

The identification and quantification in heterogeneity in disease prevalence across a range provide a geographical range and scope for targeting prevention and treatment interventions at high-prevalence or high-risk areas (Hay and Snow 2006).

Spatio–temporal dynamic relations between health measures is an imperative component in epidemiological and public health study (Nakhapakorn and Jirakajohnkool 2006). In this study, we used several techniques under the umbrella of geographic information technology (GIS). The use of a GIS with spatial statistics, including spatial filtering (smoothing) and cluster analysis, has been pertained to supplementary diseases, in which it is often used to scrutinize and more clearly display the spatial patterns of disease (Wu et al. 2009; Bonet et al. 2007). Furthermore, the recognition and quantification of heterogeneity in disease prevalence across a geographical range provides scope for targeting prevention and treatment interventions in high-prevalence or high-risk areas (Hay and Snow 2006). Spatial statistics are the most useful tool for describing and analyzing how various geographical events occur (Lee and Wong 2001).

The majority of visceral leishmaniasis (VL) cases (>90%) occurs in only six countries: India, Bangladesh, Nepal, Ethiopia, Sudan and Brazil (Chappuis et al. 2007). In the Indian sub-continent, about 200 million populations are measured to be at risk of developing VL; this region harbors an estimated 67% of the global VL disease burden. The Bihar state only has captured almost 50% cases out of total cases in Indian sub-continent (Joshi et al. 2008). There is evidence that the disease are persisting at Muzaffarpur district (Bihar), in India for many years, while others are more sporadic and persists for a short period of time only (Malaviya et al. 2011). However, more attention has been directed to vector borne disease where individual parasites exploit many hosts individual and consequently can transport a pathogen from an infected to an uninfected host. Kala-Azar or VL is well suited to cluster analysis, which intends to demarcate hotspots of high disease prevalence. The patterns of the distribution of kala-azar cases have been explored in some detail, with evidence of clustering of kala-azar around endemic areas.

In Muzaffarpur district, 15,905 cases and 127 deaths have been recorded from the period between 2007 and 2011. Conversely, the number of VL cases in Muzaffarpur district was 3,251 in 2005, fluctuated reaching a maximum of 4,920 cases in 2007, and declined reaching 2,531 cases in 2011. The VL incidence rate of the Muzaffarpur district was higher in 2007 (11.41/10,000 population, while cause-specific mortality rate was also high (4.1%) in this year. However, the case fatality rate was maximum (0.95) in 2008 in the district, during the study period. Additionally, it was also found that the lowest incidence rate of Muzaffarpur district was (5.15/10,000 population) in 2009, and case fatality rate (0.47%), followed by cause-specific mortality rate (1.1%). The districts accounted for >10% of total kala-azar cases evidenced from Bihar, and disease propagation in the district appear to be the major focus fueling a sustained epidemic. Quantitative analysis of the spatial patterns of kala-azar at the village level would enable us to identify if the observed clustering is likely to occur due to chance alone, and would help us develop hypotheses about risk factors that are associated with visceral leishmaniasis or kala-azar. Hence, in this study, we aimed to examine the spatial-temporal patterns and distribution of kala-azar disease in Muzaffarpur district of Bihar, India is using GIS tool and geo-statistical analysis.

4.3.2 *Material and Methodology*

4.3.2.1 Study Area

Muzaffarpur district lies in North Bihar plains (i.e. North of Ganga). It covers a geographical area of 3,132 Km^2 and falls under 72F, 72G and 72B degree sheets of Survey of India. It extends between North Latitude 25° 54′ 00″ to 26° 23′ 00″ and East Longitude 84°53′00″ to 85°45′00″ (Fig. 4.7). The district experiences a severe winter followed by a very hot summer (40 °C) and then by a heavy downpour of monsoon. The temperature ranges between 27 and 38 °C in general but sometimes it goes up to 44 °C also. The district receives an average rainfall of 1280 mm (CGWB 2008). The district is mainly drained by rivers Burhi Gandak, Baghmati, Baya which generally flow in a south easterly direction. As per 2011 Census (Census of India, 2011), Muzaffarpur had a population of 4,778,610 (6th rank in Bihar), with a population density of 1,514 people per km^2.

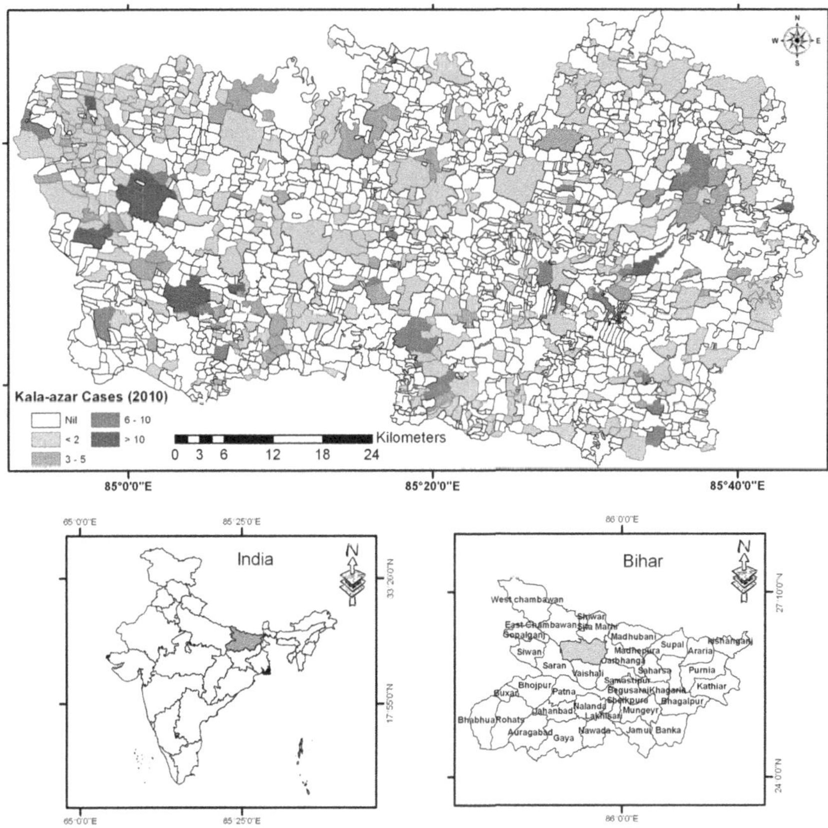

Fig. 4.7 Location map of the study area

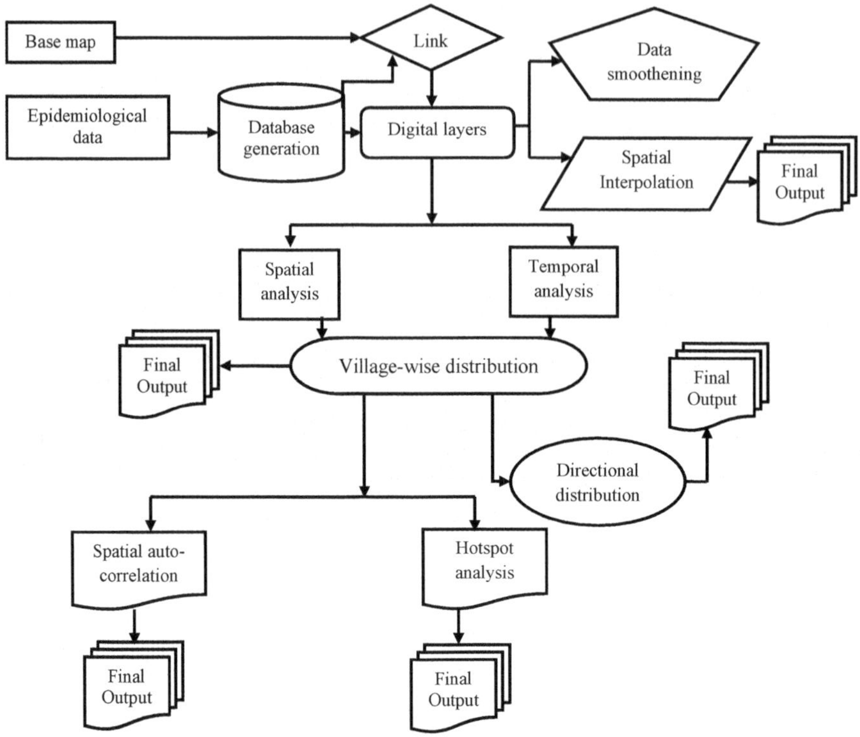

Fig. 4.8 Flow chart map of the study methodology

4.3.2.2 Data Collection and Management

Kala-azar cases data reported at village level in years 2007–2011 was used in this study. The data were acquired from the State Health Society (SHS), Bihar (India) and the Public Health Centers (PHCs). For conducting a GIS-based analysis on the spatial distribution of kala-azar, the village-level point and polygon layer was generated based on the administrative boundary map (Scale-1:25,000) obtained from the Land Revenue Office, Patna. All kala-azar cases were geo-coded and matched to the village-level layers of polygon and point by administrative code using the software ArcGIS 9.3 (ESRI Inc., Redlands, CA, USA). The flow char map of the study methodology is illustrated in Fig. 4.8.

4.3.2.3 Spatial Filtering and Smoothening

Data from all the kala-azar reported cases have been geo-coded using village location from the address of the patients. Spatial filtering can engross smoothing or sharpening the data of interest. However, the spatial smoothing was executed to lessen random noise in the data that comes from the high variance characteristic of small populations or small case numbers (Ruston and Lolonis 1996). The smoothed incidence was computed from the total number of cases per 10,000 at each village divided by the total number of people at risk within the village, which was determined using a spatial weights file including the village. Based on the annualized average incidence, all villages were clustered into four types: non-endemic area, a low endemic area with an annualized average incidence between 0 and 5 per 100,000, medium endemic area with the incidence between 5 and 30 per 100,000, and high endemic area with the incidence over 30 per 100,000. The technique of producing a smoothed map of disease rates allows for the display of data at a higher geographic scale while preserving the stability of the estimated disease rates.

4.3.2.4 Directional Distribution of Disease

Initial assessment for geographical accuracy at village level revealed sufficient information to study the spatial pattern of the kala-azar, and allocated us to use the patient address as the location of infection. Standard deviation ellipses (SDE) were used to visualize and compares the major infected areas of the disease. It divulged the directional partiality in the point distribution (Mitchell 2005). These ellipses can point out whether a spatial feature demonstrates directionality, and can also evaluate the directionality of multiple features at once. The axes for each ellipse are deliberate as one standard deviation from the mean of the x- and y-coordinates in the given data set (ESRI 2009). Furthermore, the SDE parameter for each year was measured, in order to evaluate between the years the local position and the spatial extent of disease.

4.3.2.5 Inverse Distance Weighting (IDW) Interpolation

Inverse distance weighting (IDW) interpolation method was used to map the interpolated incidence rates of Kala-Azar cases athwart the state. This is due to mapping the spatial allotment of Kala-Azar disease and potential risk areas requires, producing "bulls eyes" around data locations. IDW is an exact method that enforces that the estimated values of a point is influenced more by nearby known points than those farther away. However, in this method, there is no assessment of the pre diction errors, and no assumptions required by the data. The IDW interpolation technique is commonly used in GIS programme for producing surfaces using interpolation of scatter point and has been employed in other analysis of vector borne disease (Hu et al. 2007; Woodruff et al. 2006). The general equation for the IDW method is:

$$Z_0 = \frac{\sum_{i=1}^{s} z_i \frac{1}{d_i^k}}{\sum_{i=1}^{s} \frac{1}{d_i^k}}$$

where z_0 is the estimated value at point 0, z_i is the z value at point i, d_i is the distance between point i and 0, s is the number of known points used in estimation, and k is the specified power.

IDW weights the contribution of each input point by a normalized inverse of the distance from the control point to the interpolated point. The IDW interpolation method presumes that each input point has local influence that decreases with distance. It weights the point closer to the processing points more than those farther away (Fisher et al. 1987).

4.3.2.6 Spatial Autocorrelation

Spatial autocorrelation, also called spatial association analysis was performed in the confirmed cases of kala-azar to test whether the cases were distributed randomly over space and, if not, to appraise any recognized spatial disease clusters for statistical significance (Kulldorff et al. 1997). The global Moran's Index (Moran's I) statistics was used to evaluate autocorrelation in kala-azar spatial distribution and test how villages were clustered or dispersed in space (Boots and Getis 1998). A popular measure of spatial association is *Moran's I*, which can be computed by:

$$I = \frac{\sum_{i=1}^{n} w_{ij} \sum_{j=1}^{m} w_{ij} (x_i - \overline{X}) \left(x_j - \bar{X}\right)}{s^2 \sum_{i=1}^{n} \sum_{j=1}^{m} w_{ij}}$$

where, x_i is the value at point i, x_j is the value at points i's neighbor j, w_{ij} is a coefficient, n is the number of points, and s^2 is the variance of x values with a mean of x. The coefficient w_{ij} is the weight of measuring spatial autocorrelation. Typically, w_{ij} is defined as the inverse of the distance (d) between points I and j.

The indices were evaluated by replication, and considering the imaginative location of the villages (Fang et al. 2006). The value of *Moran's I* range from −1 to +1: a value close to '0' indicates spatial randomness while a positive value indicates positive spatial autocorrelation and vice versa. We also calculated Z-score and *p*-value linked with *Moran's I* indicates the likelihood that point pattern could be a result of random chance.

4.3.2.7 Hotspot Detection and Analysis

The hotspot is defined as a condition indicating some form of clustering in a spatial distribution (Osei and Duker 2008). This has led to the use of the Getis-Ord $G_i^*(d)$, which can separate clusters of high values from cluster of low values (Getis and Ord

1992; Mitchell 2005). The general G-statistics based on a particular distance, *d*, is defined as:

$$G(d) = \frac{\sum \sum w_{ij}(d) x_i x_j}{\sum \sum x_i x_j}, i \neq j$$

where x_i is the value at location *i*, x_j is the value at location *j* if *j* is within *d* of *i*, and $w_{ij}(d)$ is the spatial weight. The weight is based on the inverse distance if *i* and *j* represents 1 and 0 for polygon.

Moreover, clusters of cases that occur randomly can also have an influence on the spread of an infectious disease. The local $G_i^*(d)$ statistics are useful for determining the spatial dependence of neighbouring observations (Ord and Getis 1995; Hinman et al. 2006). The result expresses the Z-score and *p*-value of the calculated $G_i^*(d)$, in comparison with the normal distribution of the statistics calculated by the simulation (Feser et al. 2005), to evaluate its statistical significance. In this study, adjacency is defined using Thiessen polygon continuity weight file which has been constructed based on villages that share common vertices. A cluster of high positive Z-score suggests the presence of a cluster of high values or a hotspot. On the other hand, a cluster of high negative Z scores, suggests the presence of a cluster of the low values or a cold spot.

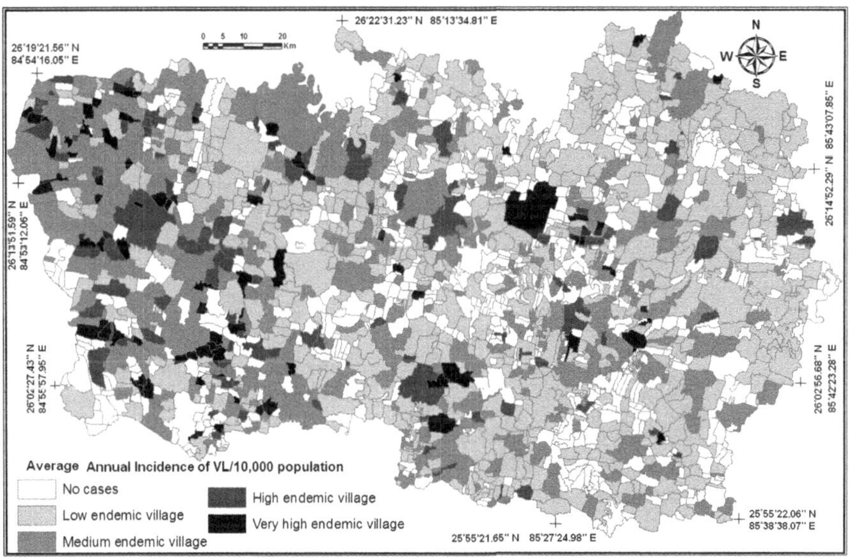

Fig. 4.9 Map showing the frequency of occurrence of kala-azar incidence from the period of 2007–2011 in Muzaffarpur district

4.3.3 Results

4.3.3.1 Spatial Distribution of Kala-Azar Cases

The distribution patterns of kala-azar incidence at the block level in Muzaffarpur district were demonstrated based on a GIS platform. The average number of kala-azar cases of the district per year is about 2,137.292. The number of notifying VL cases in Muzaffarpur district during the period from 1985 to 2011 is 66,344. A kala-azar incidence rate map of Muzaffarpur district was built from the average annual number of cases for the period 2007–2011 (Fig. 4.9), and the analysis result showed that kala-azar cases were spread all around the district, showing a very high endemic village in black colour, high endemic village in dark gray colour, a medium endemic village with lightgray colour and low endemic village with white colour. However, the results of the analysis showed most of the very high endemic villages were located in western and northwestern part of the district; some tracts were also found in the southeastern and northern regions. It was also observed that the high incidence of VL found various peripheral tracts close to expansion areas in the eastern and western regions. Nearly central and eastern part of the district showed medium and low incidence rate during the study period (2007–2011).

An interpolated layer of VL incidence rate/10,000 population was generated using the IDW method for all the consecutive year during the period from 2007 to 2011 (Fig. 4.10). Darker the areas represent the highest incidence rate, while lighter areas demarcated the low incidence rate. In 2007, the highest incidence rate was found in the southwest part of the district, while the small pockets were also delineated through the entire district. However, the low incidence rate was found in the southeastern part of the district. In 2008, the incidence rate of VL is slightly decreased. The concentration of VL is higher in the western part of the district and lower in the middle and eastern part of the study area. In 2009, some small pockets of highest incidence rate were found in the northwest and southwest part of the district.

Likewise, the low VL incidence was recorded from the eastern and southern part of the district. In 2010, the pockets of high incidence rate were extracted northern and middle part of the district. The western part of the district was documented as medium risk, while the northeast and southeast recorded the low incidence. Additionally, in 2011, the high risk zone was delineated from the western part of the district. Some small pockets of the low incidence zone were found on the eastern part of the district.

The distributional trend of VL cases in Muzaffarpur district was generated during the period from 2007 to 2011 (Table 4.2). The result pertaining to the mean centers of Muzaffarpur district shows mean centre moves towards the southern direction between 2007 and 2008, then in eastern direction between 2008 and 2009, and then towards north between 2009 and 2010 (Fig. 4.11). However, in 2011, the direction of the mean centre shifted towards the western direction again. Moreover, the magnitude of the shift between 2009 and 2010 was maximized (3.64 km).

However, in most of the PHCs, similar pattern and distribution are remaining as in the previous year. In 2011, the distribution pattern of VL is slightly dispersed in

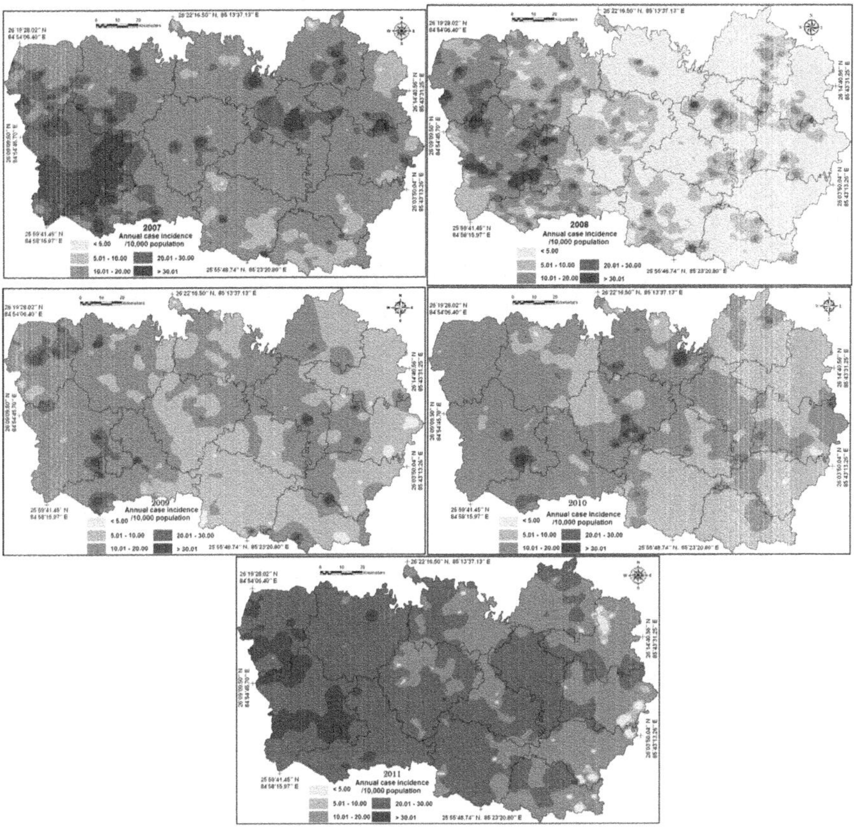

Fig. 4.10 Maps showing the inverse distance weighting interpolation incidence rates of kala-azar disease in Muzaffarpur district over different years (2007–2011)

Table 4.2 Distance between mean centre and their details over the year of kala-azar incidences of Muzaffarpur district, Bihar

Year	Distance from previous year mean centre	Standard distance (m)	X-coordinate	Y-coordinate
2007	1.65 km (2007–2008)	25185.04	321904.96	2894307.2
2008	2.64 km (2008–2009)	24899.86	322671.87	2892858.74
2009	1.66 km (2009–2010)	24378.50	325341.02	2892965.9
2010	3.74 km (2010–2011)	25582.62	326177.78	2894397.6
2011	0.79 km (2007–2011)	25049.99	322486.19	2893784.21

the western part of the district. Finally, major shifting of VL distribution has been marked out in the Kanti, Bochaha, and Minapur PHC during the period from 2007 to 2011. The analysis also illustrated that ellipses were slightly shifted from the previous year in Musahari, Gaighat, Kurhani, Sahebganj and Dholi PHCs (Fig. 4.11 and Table 4.3).

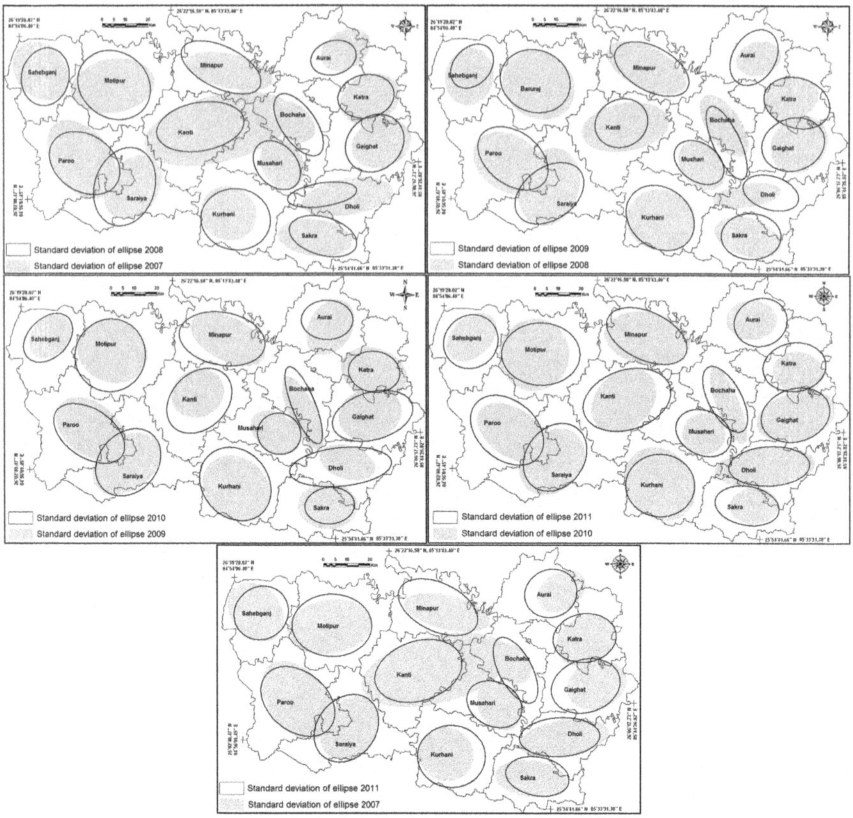

Fig. 4.11 Standard deviation of ellipse (SDE) showing the directional pattern of kala-azar distribution at sub-district level of Muzaffarpur district over the period of 2007–2011

Table 4.3 Values calculated by standard deviation of ellipse (SDE) and their details over the year of Kala-Azar incidences of Muzaffarpur district, Bihar

Year	Xstd distance (m)[a]	Ystd distance (m)[a]	Rotation (degrees)
2007	32005.05	15628.48	99.15
2008	31681.33	15372.03	97.40
2009	30574.95	15930.94	104.64
2010	32495.87	15904.71	101.89
2011	31678.27	15858.48	101.88

[a]Long axes (Xstd distance) and Short axes (Ystd distance)

4.3.3.2 Spatial Auto-correlation of Kala-Azar Distribution

Global Moran's I statistics measured the autocorrelation of the incidence ratio among all villages including villages with less VL incidence and/or no cases. The

Table 4.4 Spatial autocorrelation estimation of different years using *Moran's* in Muzaffarpur district during the period from 2007 to 2011

Year	Observed	Expected	Variance	Z-score	*P*-value	Pattern
2007	0.085841	−0.000489	0.000166	6.704012	0.000000	Clustered
2008	0.127962	−0.000489	0.000170	9.866017	0.000000	Clustered
2009	0.050241	−0.000489	0.000171	3.878790	0.000105	Clustered
2010	0.060073	−0.000489	0.000171	4.624768	0.000004	Clustered
2011	0.196729	−0.000489	0.000173	14.986595	0.000000	Clustered

spatial autocorrelation for annual incidence rate of villages in Muzaffarpur district from 2007 to 2011 showed a significant positive autocorrelation. The values of *global Moran's I* varied from 0.05 to 0.20 value with less than the 0.01 significance level for each year (Table 4.4).

However, the result of the analysis exemplified that the spatial distribution of VL affected villages were somewhat spatially auto-correlated but there is a chance of random pattern. Additionally, the highest *global Moran's I* value was recorded in 2011, while in 2009 the value of *Moran's I* was much lower. For example, in 2009, the value of global Moran's I index was 0.05 with a Z-score 3.87. Therefore based on the results of this global spatial test the hypothesis of spatial randomness could not be rejected, meaning that the spatial pattern of VL incidence was slightly random across the district.

4.3.3.3 Hotspot and Cold Spot Analysis of Kala-Azar Distribution

In the map (Fig. 4.12), the 'dark' areas indicate statistically significant ($P < 0.01$) hotspots, while, 'white' areas represent significant cold spot areas. In 2007, hotspot areas were demarcated to the western and northwestern corner of the district. In 2008, there was a sharp increase hotspot areas that were found in the western and upper central part of the district; however, some small pockets of cold spot areas were illustrated in the northern and southwestern regions. In 2009, there was a sharp decrease in VL hotspot in the western region, which showed less risk. Moreover, we observed a slight increase in VL hotspot in the central and southern regions. In 2010, most of the hotspot areas were found in the north western, eastern and south eastern region. Conversely, some small pockets were illustrated north and south-western regions. On the other hand, cold spots areas were found in the central part of the study area, while, these are found in the urban and peri-urban region. In 2011, we observed VL hotspot foci in the northwestern and western region of the district, and some small pockets in northeastern region. However, cold spot areas were segregated in the central and eastern and south eastern corner of the district. Finally, a hotspot region was delineated of the Muzaffarpur district during the period from 2007 to 2011, showed western part of the district is highly affected by kala-azar compare to the eastern part. And most of the hotspot regions are found in the

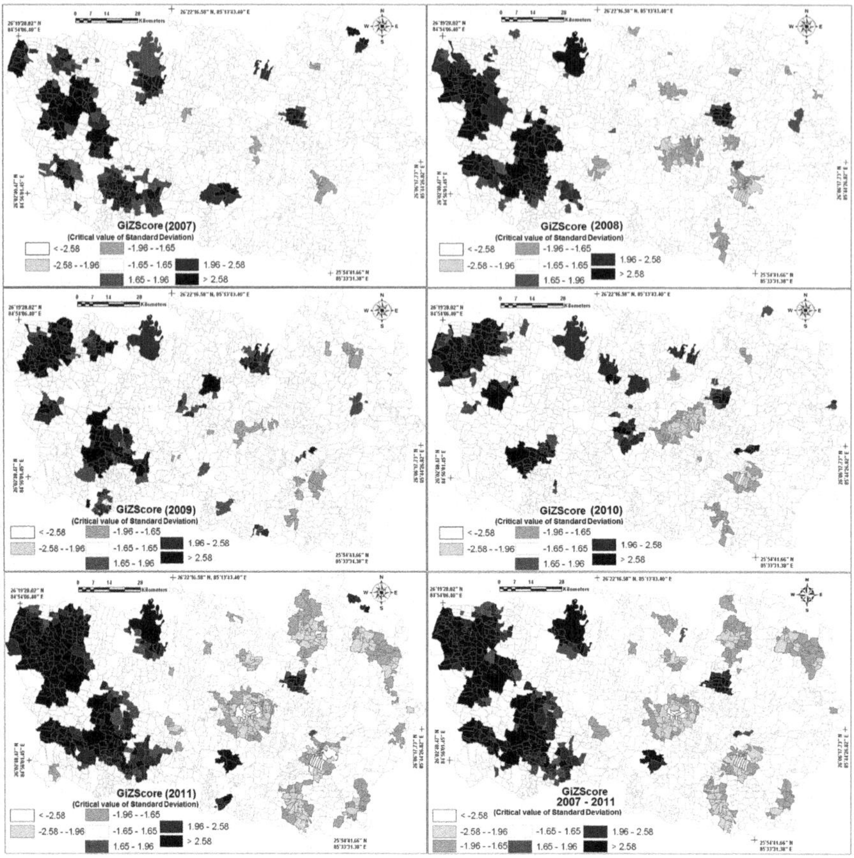

Fig. 4.12 Maps showing the hotspot and cold spot over different years (2007–2011) in Muzaffarpur district

western and small pockets of central and northern part of the study area. Additionally, cold spot areas are also portrayed on the eastern and central part of the study area.

4.3.4 Discussion

The Muzaffarpur district has the highest reported number of VL cases in Bihar during the last two-three decades. Analysis of the incidence of kala-azar across the entire Muzaffarpur district has never been comprehensively assessed, although the disease incidence has been compared between selected locations. However, this is the first regional examination of the incidence of VL throughout the Muzaffarpur district over a period of five consecutive years (2007–2011) at village level. The investigation has been carried out to delineate the spatial and temporal distribution of VL and provide an appropriate method for identifying the epidemic in local areas. The endorsement of public health facilities for VL treatment was started since 2006 when monetary incentives were provided to patients attending PHCs in Bihar. It is noteworthy that the underreporting of VL decreased from 2003 to 2006 in Bihar (Singh et al. 2010) and the number of cases reported by NGOs increased during this period. Furthermore, the spraying techniques and upgrading of household characteristics, concentration should be focused on the primary health care approach to vector control together with the involvement of the community and the health service infrastructure. Usefulness of environmental management with modification aiming at reducing and controlling VL proved to be more efficient than the application of pesticides (Warburg and Faiman 2011).

This study reveals the spatial and temporal characteristics of kala-azar disease in Muzaffarpur district of Bihar (India) using GIS tools and geostatistical analysis, which allow for the quantification of the degree of clustering of infections. However, to our knowledge this is the first attempt to implement GIS mapping techniques to examine the distribution pattern of kala-azar at micro-level in Muzaffarpur district of Bihar (India). This study has three foremost strengths. Initially, this is the first study to inspect the geographic variation of kala-azar disease across geopolitical borders in India based on spatial statistics. This study lays a foundation for a further investigation of the spatial and temporal patterns and the risk factors of this disease. Secondly, the results of this study demonstrate that GIS mapping techniques may be used as a tool to quickly display information and generate maps to highlight kala-azar disease risk prone areas for developing more effective control and prevention strategies. The maps could be used to suggest high risk areas where further investigation should be focused, to identify whether increased disease surveillance measures or possible control activities are warranted. Third, the kala-azar disease data used in this study are somewhat comprehensive, may be used in the national level. Finally, in this study, we aimed to examine the distribution patterns of kala-azar disease spatially and temporarily at the smallest geographical unit of Bihar, India.

The standard deviation of ellipse not only describes VL spread direction. The analysis of SDEs is indicated that the overall spread tendency in Muzaffarpur district of the western part. The analysis of the results is also corroborated with previous studies (Malaviya et al. 2011). Although the distributed center and the range are clear, the degree of the concentration or dispersed is not precise.

The present result is indicated that geostatistics approach if carefully applied can play an important role in the recognition and analysis of the spatial structure for kala-azar epidemiology and control. The spatial association between cases is subject to the measure of geographical "closeness" or spatial proximity rather than a formal analysis (White 2005). The study showed that spatial distribution patterns of kala-azar cases were significantly clustered, and identified the kala-azar hotspots in Muzaffarpur district. In this study, a strong evidence of spatial autocorrelation of kala-azar disease is observed across the district using Moran's I statistics. The distribution pattern of VL in Muzaffarpur district is clustered, but there is a chance of random pattern. The highest Moran's I value of Muzaffarpur district was recorded in 2011 (*Moran's I*–0.20), illustrated that the spatial autocorrelation of VL affected villages is stronger compared to the years of 2007–2010. The study has been carried out with significance level of $p < 0.01$. However, cluster analysis did not show any clear spread pattern or trend. This could be due to the multiple factors. It may be due to the fact that the VL or kala-azar transmission is going on unabatedly (e.g., shimmering transmission). Additionally, heterogeneity in the revival time of unhygienic individuals and behavioral transformations, persuaded by the occurrence of cases, would modify the experimental spatial pattern.

Getis-Ord G_i^* statistic was used to identify the kala-azar hotspot and cold spot based on the incidence rate/10,000 population in each year for the study districts. The analysis of the result suggests that the disease is spreading nearby foci, with waves of attentiveness dissemination process of hotspot and cold spot. The results revealed that cold spot was found in the urban and peri-urban region for both the study area. This may be due to the several conditions that favour sand fly density, local climate, housing condition etc. This information could help to epidemiologist and health management professionals to mark out the areas of good habitat of vector and susceptible population for kala-azar.

Some limitation of this study should also be painted, viz. although the comparatively high coverage of geo-coding of VL cases; it is likely to have been worse in peripheral areas of the district. The situation may have caused distortions in the estimates, possibly leading to the either spurious identification of high-risk areas and non-identification of specific foci. Second, whole analysis has been performed based on the area data. Notwithstanding these constraints, the study illustrated GIS and GIS-based spatial statistical techniques may provide an opportunity to clarify and quantify the epidemic situation of kala-azar within re-emerged epidemic areas, and put down an underpinning to chase future investigations into the environmental factors responsible for the increased disease jeopardize. Additionally, the methodology is based on a philosophy on general principles of geostatistics may employ the model to plan a strategy to control kala-azar by the information received for distribution and hotspots for various months or seasonally.

4.4 Conclusion

The 'state of art' in statistical techniques suitable to definite problems in spatial epidemiology comprises some powerful, multitalented and valuable tools. Many of the existing spatial approaches and models are practically widely known in the statistical community and some of them have been in employ for several years. Valuable spatial epidemiology does not essentially chase from the employ of better and more complicated statistical methods. Clearly, the critical remuneration of the entire statistical attempt in spatial epidemiology significantly depends on appropriate and well founded epidemiological contemplations united with admittance to data at a proper level of detail and of adequate quality to deal with the topic under consideration. Ultimately, environmental epidemiology desires to be assessed in the same way as any other public health screening programs. Such applications determine the spatial statistical investigation of already accessible data or that begin from routine collection systems.

References

Anselin L (1995) Local indicators of spatial association—LISA. Geogr Anal 27:93–115

Auchincloss AH, Gebreab SY, Mair C, Roux AVD (2012) A review of spatial methods in epidemiology, 2000–2010. Annu Rev Public Health 33:107–122. https://doi.org/10.1146/annurev-publhealth-031811-124655

Bailey TC (2001) Spatial statistical methods in health. Cad. Saúde Pública, Rio de Janeiro, 17(5): 1083–1098

Balakrishnan P, Saleem A, Mallikarjun ND (2011) Groundwater quality mapping using geographic information system (GIS): a case study of Gulbarga City. Karnataka, India. 5(12): 1069–1084

Bhat S, Motz LH, Pathak C, Kuebler L (2015) Geostatistics-based groundwater-level monitoring network design and its application to the Upper Floridan aquifer, USA. Environ Monit Assess 187:4183

Bhunia GS, Kesari K, Chatterjee N, Kumar V, Das P (2013) Spatial and temporal variation and hotspot detection of kala-azar disease in Vaishali district (Bihar), India. BMC Infect Dis 13:64. https://doi.org/10.1186/1471-2334-13-64

Bhunia GS, Shit PK, Maiti R (2016) Comparison of GIS-based interpolation methods for spatial distribution of soil organic carbon (SOC). J Saudi Soc Agric Sci. http://dx.doi.org/10.1016/j.jssas.2016.02.001

Bindiya MV, Unnikrishnan A, Poulose JK (2013) Spatial clustering algorithms—an overview. Asian J Comput Sci Informat Technol 3:1–8

Bonet M, Spiegel JM, Ibarra AM, Kouri G, Pintre A, Yassi A (2007) An integrated ecosystem approach for sustainable prevention and control of dengue in Central Havana. Int J Occup Environ Health 13:188–194

Boots BN, Getis A (1998) Point pattern analysis Newbury Park. Sage Publications, Newbury Park, CA, USA

Caers J (2005) Petroleum geostatistics. An SPE Primer, Society of Petroleum Engineers, Richardson, TX, USA

Central Ground Water Board (CGWB) (2008) Ground water information booklet—Muzaffarpur district, Bihar state. Ministry of Water Resources, (Govt. of India), Mid-Eastern Region, Patna. http://cgwb.gov.in/District_Profile/Bihar/Muzaffarpur.pdf

Chappuis F, Sundar S, Hailu A, Ghalib H, Rijal S, Peeling RW, Alvar J, Boelaert M (2007) Visceral leishmaniasis: what are the needs for diagnosis, treatment and control? Nat Rev (Microbiology) 5:873–882

Cliff AD, Haggett P (1988) Atlas of disease distributions. Oxford, United Kingdom: Blackwell

Cliff AD, Ord JK (1973) Spatial autocorrelation. Pion, London

Cliff AD, Ord JK (1969) The Problem of Spatial Autocorrelation. In: Scott AJ (ed) Studies in regional science London papers in regional science, pp 25–55. Pion, London

Cressie N (1993) Statistics for spatial data, rev.edn. Wiley, New York, NY (pp 900). (Original edition, 1991. Paperback edition in the Wiley Classics Library: Wiley, Hoboken, NJ, 2015)

de la Cruz ML, Perez A, Bezos J, Pages E, Casal C, Carpintero J et al (2014) Spatial dynamics of bovine tuberculosis in the autonomous community of Madrid, Spain (2010–2012). PLoS One 9(12):e115632

Deutsch CV, Journel AG (1992) Geostatistical software library and user's guide. Oxford University Press, p 340

Diaz-Lacava AN, Walier M, Holler D, Steffens M, Gieger C, Furlanello C, Lamina C, Wichmann HE, Becker T (2015) Genetic geostatistical framework for spatial analysis of fine-scale genetic heterogeneity in modern populations: results from the KORA study. Int J Genom, Article ID 693193, p 15. https://doi.org/10.1155/2015/693193

Dionissios H (2015) Local geostatistical models and big data in hydrological and ecological applications. EGU General Assembly held 12–17 April, 2015 in Vienna, Austria. ID 2179

Ebdon D (1985) Statistics in geography. Blackwell

Elliott P, Wakefield J, Best N, Briggs D (2000) Spatial epidemiology: methods and applications, Oxford University Press

Enkhtur B (2013) Geostatistical modelling and mapping of air pollution. https://www.itc.nl/library/papers_2013/msc/gfm/enkhtur.pdf

Environmental Systems Research Institute (ESRI), Inc (2009) How directional distribution: standard deviational ellipse (spatial statistics) works. http://webhelp.esri.com/arcgiSDEsktop/9.3/index.cfm? TopicName5—How directional distribution: standard deviational ellipse (Spatial Statistics) works

ESRI (2001) Using ArcGIS geostatistical analyst. ESRI Press, Redlands, CA

Fang L, Yan L, Liang S, Vlas SJD, Feng D, Han X, Zhao W, Xu B, Bian L, Yang H, Gong P, Richardus JH, Cao W (2006) Spatial analysis of hemorrhagic fever with renal syndrome in China. BMC Infect Dis 6:77

Feser E, Sweeney S, Renski H (2005) A descriptive analysis of discrete U.S. industrial complexes. J Regional Sci 45:395–419

Few S (2009) Introduction to geographical data visualization. Perceptual Edge Visual Business Intelligence Newsletter. https://www.perceptualedge.com/articles/-visual_business_intelligence/geographical_data_visualization.pdf

Fisher NI, Lewis T, Embleton BJJ (1987) Statistical analysis of spherical data. Cambridge Unviersity Press, Cambridge

Fouedjio F (2016) Space Deformation Non-stationary geostatistical approach for prediction of geological objects: Case Study at El Teniente Mine (Chile). Nat Resour Res 25:283. https://doi.org/10.1007/s11053-015-9287-7

Friendly M (2008) Milestones in the history of thematic cartography, statisticalgraphics, and data visualization. http://www.math.usu.edu/ ~ symanzik/teaching/2009_stat6560-/Downloads/Friendly_milestone.pdf

Gandin LS (1963) Objective analysis of meteorological fields, Leningrad, Gidrometeorologicheskoe Izdatel'stvo (GIMIZ)

Geary R (1954) The contiguity ratio and statistical mapping. Inc Stat 5(3):115–145

Georges Matheron (1963) Principles of geostatistics. Econ Geol 58(8):1246–1266

Getis A (2008) A history of the concept of spatial autocorrelation: a geographer's perspective. Geogr Anal 40:297–309

Getis A, Ord JK (1992) The analysis of spatial association by use of distance statistics. Geogr Anal 24:189–206

Goovaerts P (2006) Geostatistical analysis of disease data: visualization and propagation of spatial uncertainty in cancer mortality risk using poisson Kriging and P-Field simulation. Int J Health Geogr 5:7

Griffith D (1987) Spatial autocorrelation: a primer. Resource Publications in Geography, Association of American Geographers

Griffith DA (2009) Spatial autocorrelation. http://booksite.elsevier.com/-brochures/hugy/SampleContent/Spatial-Autocorrelation.pdf

Gunarathna MHJP, Kumari MKN, Nirmanee KGS (2016) Evaluation of interpolation methods for mapping pH of groundwater. IJLTEMAS V(III):1–5

Han J, Kamber M (2006) Data mining: concepts and techniques, 2nd edition, Morgan Kaufmann, San Francisco, CA

Hani A, Abari SAH (2011) Determination of Cd, Zn, K, pH, TNV, organic material and electrical conductivity (EC) distribution inagricultural soils using geostatistics and GIS (case study: SouthWestern of Natanz—Iran). World Acad Sci Eng Technol 5(12):22–25

Hart JF (1954) Central tendency in areal distributions. Econ Geogr 30:48–59

Hay SI, Randolph SE, Rogers DJ (2002) Advances in parasitology. Academic Press, London, Remote Sensing and Geographical Information system in Epidemiology. ISBN 0-12-333560-4

Hay SI, Snow RW (2006) The malaria atlas project: developing global maps of malaria risk. PLoS Med 3:e473

Hinman S, Blackburn JK, Curtis A (2006) Spatial and temporal structure of typhoid outbreaks in Washington DC., 1906–1909: evaluating local clustering with the *i G statistic. Int J Health Geogr 5:13

Holliger K, Tronicke J, Paasche H, Dafflon B (2008) Quantitative integration of hydrogeophysical and hydrological data: geostatistical approaches. Overexploitation and contamination of shared groundwater resources. Springer Netherlands, ISBN 978-1-4020-6985-7. https://doi.org/10.1007/978-1-4020-6985-7_5

Hu W, Tong S, Mengersen K, Oldenburg B (2007) Exploratory analysis of social and environmental factors associated with the incidence of Ross River viruses in Brisben, Australia. Am J Trop Med Hyg 76:814–819

Hubert LJ, Golledge RG, Costanza CM (1981) Generalized procedures for evaluating spatial autocorrelation. Geogr Anal 13:224–232

Isaaks EH, Srivastava M (1989) An introduction to applied geostatistics. Oxford University Press, New York

Jensen OP, Christman MC, Miller TJ, Jensen AF, Olaf P, Christman MC, Miller TJ (2006) Landscape-based geostatistics: a case study of the distribution of blue crab in Chesapeake Bay. Environmetrics 17(6):605–621

Johnston K, Ver Hoef JM, Krivoruchko K, Lucas N (2001) Using ArcGIS geostatistical analyst. ESRI Press, Redlands, CA

Joshi A, Narain JP, Prasittisuk C, Bhatia R, Hashim G, Jorge A, Banjara M, Kroeger A (2008) Can visceral leishmaniasis be eliminated from Asia? J Vector Borne Dis 45:105–111

Journel AG (1986) Geostatistics: models and tools for the earth sciences. Math Geol 18:119–140. https://doi.org/10.1007/BF00897658

Journel AG, Huijbregts CJ (1978) Mining geostatistics. Academic Press, London, p 600

Kerry R, Goovaerts P, Haining RP, Ceccato V (2010) Applying geostatistical analysis to crime data: car-related thefts in the Baltic States. Geogr Anal 42(1):53–77. https://doi.org/10.1111/j.1538-4632.2010.00782.x

Krivoruchko K, Butler K (2013) Unequal probability-based spatial mapping. Esri, Redlands, CA, USA. http://www.esri.com/esrinews/arcuser/spring2013/ ~ /-media/Files/Pdfs/news/arcuser/0313/unequal.pdf

Kulldorff M, Feuer EJ, Freedman LS (1997) Breast cancer clusters in the Northeast United States: a geographic analysis. Am J Epidemiol 146(2):161–170

Kumar A, Kadiyala A, Sarmah D (2014) Evaluation of geographic information systems-based spatial interpolation methods using Ohio Indoor Radon data. Open Environ Eng J 7:1–9

Lawson AB (1989) Score tests for detection of spatial trend in morbidity data. Dundee, Dundee Institute of Technology

Lee J, Wong DWS (2001) Statistical analysis with ArcView GIS. Wiley, New York

Li XF, Chen ZB, Chen HB, Chen ZQ (2011) Spatial distribution of soil nutrients and their response to land use in eroded area of South China. Proc Environ Sci 10:14–19

Maingi JK, Mukeka JM, Kyale DM, Muasya RM (2012) Spatio-temporal patterns of elephant poaching in south-eastern Kenya. Wildlife Res 39(3):234–249

Malaviya P, Picado A, Singh SP, Hasker E, Singh RP, Boelaert M, Sundar S (2011) Visceral leishmaniasis in Muzaffarpur district, Bihar, India from 1990 to 2008. PLoS One 46(3):e14751

Mitchell A (2005) The ESRI Guide to GIS Analysis: Volume 2 Spatial Measurements and Statistics. ESRI Press, Redlands, California

Moran PAP (1950) Notes on continuous stochastic phenomena. Biometrika 37:17–23

Moran PAP (1948) The interpretation of statistical maps. J Roy Stat Soc B 10:243–251

Nakhapakorn K, Jirakajohnkool S (2006) Temporal and spatial autocorrelation statistics of dengue fever. Dengue Bulletin 30:177–183

Nalder IA, Wein RW (1998) Spatial interpolation of climatic normals: test of a new method in the Canadian boreal forest. Agric For Meteorol 92:211–225

Olea RA (1991) Geostatistical glossary and multilingual dictionary. Oxford University Press, New York

Openshaw S, Charlton M, Wymer C, Craft A (1987) A mark 1 geographical snalysis machine for the automated analysis of point data sets. Int J Geogr Inf Syst 1(4):335–358

Ord JK, Getis A (1995) Local spatial autocorrelations statistics: distributional issues and application. Geogr Anal 27:286–306

Osei FB, Duker AA (2008) Spatial and demographic patterns of Cholera in Ashanti region—Ghana. Int J Health Geogr 7:44

Pilz J, Spöck G (2007) Why do we need and how should we implement Bayesian Kriging methods. Stoch Env Res Risk Assess 22(5):621–632

Relethford JH (2008) Geostatistics and spatial analysis in biological anthropology. Am J Phys Anthropol 136:1–10

Rushworth AM, Peterson EE, Ver Hoef JM, Bowman AW (2015) Validation and comparison of geostatistical and spline models for spatial stream networks. Environmetrics 26(5):327–338

Ruston R, Lolonis P (1996) Exploratory spatial analysis of birth defect rates in an urban population. Stat Med 15:717–726

Sarah EB, Daniel S, Michael PF (2016) Shapes on a plane: evaluating the impact of projection distortion on spatial binning. Cartogr Geogr Informat Sci 1–12. https://doi.org/10.1080/15230406.2016.1180263

Shit PK, Bhunia GS, Maiti R (2016) Spatial analysis of soil properties using GIS based geostatistics models. Model Earth Syst Environ 2:107

Shyti B, Fetahu E (2015) Spatial statistical methods in the analysis of public health data. Eur Scientif J 11(21):47–55

Singh VP, Ranjan A, Topno RK, Verma RB, Siddique NA, Ravidas VN, Kumar N, Pandey K, Das P (2010) Estimation of under-reporting of visceral leishmaniasis cases in Bihar. India. Am J Trop Med Hyg 82(1):9–11

Skøien JO, Merz R, Bloschl G (2006) Top-kriging—geostatistics on stream networks. Hydrol Earth Syst Sci 10:277–287

Stein ML (1999) Predicting random fields with increasingly dense observations. Ann Appl Probab 9:242–273

Tobler WR (1970) A computer movie simulating urban growth in the detroit region. Econ Geogr 46:234

Tola E, Al-Gaadia KA, Madugundua R, Zeyadaa AM, Kayad AG, Biradar CM (2016) Characterization of spatial variability of soil physicochemical properties and its impact on Rhodes grass productivity. Saudi J Biol Sci. http://dx.doi.org/10.1016/j.sjbs.2016.04.013

Waller LA, Jacquez GM (1995) Disease models implicit in statistical tests of disease clustering. Epidemiology 6:584–590

Warburg A, Faiman R (2011) Research priorities for the control of phlebotomine sand flies. J Vector Ecol 36(1):S10–S16

White P (2005) Spatial data cluster analysis. Public Health Intelligen, Ministry of Health New Zealand

Whittle P (1954) On stationary processes in the plane. Biometrika 41:434–449

Woodruff RE, Guest GS, Garner MG, Becker N, Lindsay M (2006) Early warning of Ross River virus epidemics: combining surveillance data on climate and mosquitoes. Epidemiology 17:569–575

Wu CF, Lai CH, Chu HJ, Lin WH (2011) Evaluating and mapping of spatial air ion quality patterns in a residential garden using a geostatistic method. Int J Environ Res Public Health 8(6):2304–2319

Wu PC, Lay JG, Guo HR, Lin CY, Lung SC, Su HJ (2009) Higher temperature and urbanization affect the spatial patterns of dengue fever transmission in subtropical Taiwan. Sci Total Environ 407:2224–2233

Youden WJ (1951) Statistical methods for chemists. John Wiley & Sons, New York

Zeqiri R, Kelmendi S, Zeqiri I (2012) Geostatistics in modern mining planning. J Int Environ Appl Sci 7(2):310–317

Chapter 5
Exploring Ecology and Associated Disease Pattern

5.1 Introduction

The U.S. Centers for Disease Control (CDC) and Prevention describes emerging diseases as "*diseases of infectious origin whose incidence in humans has increased within the past two decades or threatens to increase in the near future*" (Centers for Disease Control and Prevention 1994). The World Health Organization has stated that emerging infections "*represent a global threat that will require a coordinated global response*" (WHO 1995). Approximately 1,400 species of infectious agents have been documented to cause disease in humans (Woolhouse and Gowtage-Sequeria 2005). Morse and Schluederberg (1990) suggested that emerging infectious disease (EID) refers to "*the diseases new to a population or that has recently increased in incidence or geographic range*", and could be feast in an entire region within a few days (Morse 1995). For example, in 1980s, acquired immune deficiency syndrome (AIDS), caused by the human immunodeficiency virus (HIV) emerged and disseminated briskly throughout the United States (Gould 1993). The CDC estimates 1,051,875 cases and 583,298 deaths have ensued since the epidemic began (CDC 2009). Between 17th and early 20th centuries, vector-borne diseases were liable for more human deaths than all other causes united (Gubler 1991). Approximately 3,500 different species and sub-species, mosquitoes are extremely copious and able to acclimate to closely every type of habitat (Eldridge 2005).

Globalization, climate change, unstable weather condition, pathogen evaluations, drug and insecticide confrontation, growing vector habitat and populations, and the deficiency of preventive measures have eradicated barriers and fortified the emergence and reemergence of vector-borne diseases during the course of the world (Ozer 2005). The present biodiversity crunch is comparatively due to the appearance of diseases that instigated in domestic animals or in man. As a consequence of global change, wildlife, livestock and man are fixed to inherit interaction progressively and frequently, hence growing the risk of transmission of zoonoses and

G. S. Bhunia and P. K. Shit, *Geospatial Analysis of Public Health*,
https://doi.org/10.1007/978-3-030-01680-7_5

zooanthroponoses (Hudson et al. 2006). There are a sequence of various ecological disseminations or baseline risk map available which have been resultant for a variety of tenacities, by a extensive community of public health cartographers by a dissimilar toolbox of mapping methods (Stevens and Pfeiffer 2011; Pfeiffer et al. 2008). There are varieties of causes for inadequate to map the geographical distribution of an infectious disease. Maps of disease propagation and amount countenance an instant picturing of the extent and magnitude of the public health problem (Hay et al. 2013).

5.2 Challenges of Emerging Infectious Disease

Emerging infectious diseases are diseases of infectious origin whose incidence in humans has grown within the current past or impends to rise in the near prospect (Morse and Schluederberg 1990). Approximately 15 million (>25%) of 57 million annual deaths globally are assessed to be connected openly to infectious diseases; this number does not contain the supplementary millions of deaths that happen as a magnitude if past infections, or because of the difficulties associated with the chronic diseases (Morens et al. 2004). These also comprise those contagions that seem in different geographic areas or proliferate sharply (Fig. 5.1). The diseases are re-emerging after phases of dormancy are also assembled under emerging infectious diseases. Emerging infectious diseases portend to interrupt the health care system. The prolific appreciation and usage of emerging infectious diseases is becoming a

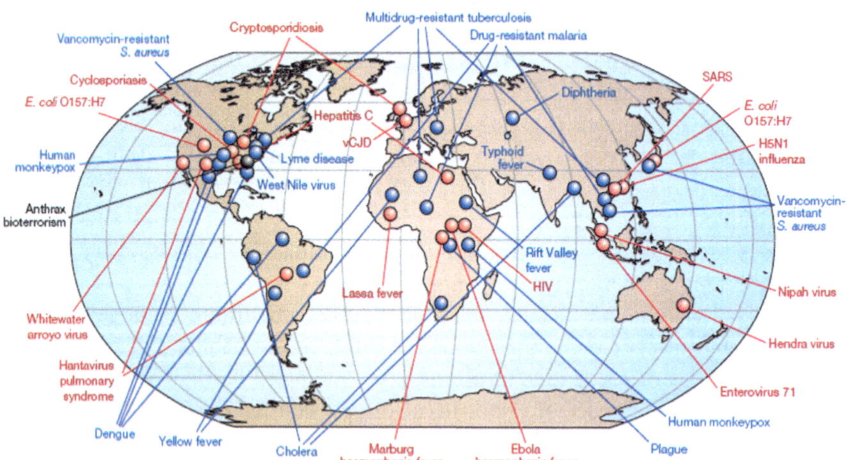

Fig. 5.1 Global examples of emerging and re-emerging infectious diseases. Red represents newly emerging diseases; blue, re-emerging/resurging diseases; black, a 'deliberately emerging' disease (*Sources* Fauci, A. S. Infectious diseases: considerations for the 21st century. Clin. Infect. Dis. 32, 675–685 (2001))

hostile task. The intermittent incidence of epidemics of infectious diseases assists to emphasize the significance of the public health system. To conflict emergent infective diseases, public health, consequently, desires to fortify and inflate investigation on the epidemiology and ecology of microbes, vectors and intermediary hosts, and generate consciousness on the prospect that new epidemics can, will appear in unanticipated spaces.

The encumbrance of morbidity and mortality related with infectious diseases descent most profoundly on people in developing countries (Guerrant and Blackwood 1999), and predominantly on infants and children. In advanced countries, infectious disease mortality disproportionately disturbs indigenous and neglected minorities (Butler et al. 2001). Infectious diseases endure to be main challenge in the South-East Asia. It is assessed that about 40% of the 14 million deaths per annum in the region and account for 28% of the worldwide encumbrance of infectious disease. The region has seeming plentiful epidemics of new and incipient infections as new micro-organisms endure to execute and certain prevailing ones amend their physiognomies to endorse their existence at the period spread of human health. Ebola virus, Zika virus, Japanese encephalitis, Nipahvirus, leptospirosism Chandipora virus are instances of emergent infectious disease that seemed a few years back and have now recognized endemicity. New pathogens endure erratic and remain to perform and spread across countries, without concerning national borderlines. They tolerate to happenstance our competence to squelch to the epidemic quickly.

The contest of emerging infection is here and will persist in the predictable prospect. The extent and antimicrobial confrontation calamity is indefinite because of the nonappearance of regular observation in developing countries. The virus may not be demolished by the conformist antiviral drugs and the accessible vaccine incapable to deliberate resistance against it. Recognition of new virus shall also depend upon the efficacy of worldwide reconnaissance survey and competences of the hitherto mistreated public health research laboratory. Hence, the susceptibility of the human race against emerging infections is dominant; the armaments with microbes immeasurable. Though there have been noteworthy attainments on numerous fronts, a proportion of other to be done to contest embryonic infectious malady.

5.3 Ecological Conditions and Disease Interaction

Infectious diseases threatened individual happiness is a major public health burden in developing countries. The world Health Organization (WHO) reports yearly on the numbers of deaths and disability adjusted life years (DALYs) by infection group in different parts of the biosphere (WHO 2009). The human ecology of disease is hesitant with the habits of social comportments, in its cultural and socio-economic circumstances, intermingles with environmental surroundings to produce or avert disease among predisposed people. Geographers have usuallyconsidered the formation of landscape; the movement and composition of population (De Sherbinin et al. 2007), factors of economic activity (Adda 2015) and its place and the dissemination of things, ideas and technology. Till date, very less number of studies is available regarding the significance of behavioral and sociocultural factors in influential patterns of disease transmission at finer spatial scales.

The key event—the 'environmental' episode that commences a fresh human infection—is a novel physical contact between possible pathogen and human (McMichael 2004). The infectious agent regularly develops from an animal source, although some origin from the soil. A scrupulous constraint, usually, is that the latent pathogen is a transformed strain that casually has become enhanced able to penetrate and live in the human host. This contact event may occur naturally. More frequently, it emerges that such contacts have happened due to some cultural, social, behavioral or technological change on the part of humans. The succeeding extend of the 'new' infectious diseases may depend on either environmental or social factors.

These include the following:

- Demographic characteristics and processes, human mobility, etc.;
- Land use, other environmental changes, encroachment on new environments;
- Consumption behaviors (eating, drinking, and, more generally, culinary culture);
- Other behaviors (sexual contacts, IV drug use, hospital procedures, etc.);
- Host condition (malnutrition, diabetes, immune status, etc.).

An ecosystem perspective on the risks to humans from vector borne disease and its control comprises an obligation of the role joked by main environmental vicissitudes and by native ecosystems in supporting vector proliferations and empowering disease transmission (McMichael 2004). The human induced undertakings (deforestation, settlements, development of dam, agricultural activities etc.) that amend the natural ecosystems also interrupt the transmission cycles of vector borne infectious diseases. These variations may clue to the fading of natural balance, amended ambiances for disease propagation and disease epidemic to humans.

Variations to climatic, ecological, demographic and social circumstances can all promote infectious diseases in various ways. Ecological aspects habitually precipitate occurrence by retaining people in contact with a natural reservoir or host for an contagion before unacquainted but typically existent either by increasing immediacy or by changing circumstances so as to errand an increased population of the microbe or its natural host (Raulerson 2010). A determinant is several aspects or variable that can influence the frequency with which a disease ensues in a population. Determinants can be generally categorized as being either intrinsic or extrinsic in nature. Intrinsic determinants are physical or physiological features of the host or disease agent which are usually dogged genetically. Extrinsic factors are typically concomitant with some practice of environmental effect on the host or disease agent. The extrinsic factors also influence the any transitional hosts or vectors convoluted in the spread of a disease, and consequently control the form and magnitude of the disease transmission taking place. They may also contain interferences made by man into the disease progression by practice of drugs, vaccines, dips, program controls and seclusions.

In case of infectious diseases, the existence or non-existence of the aetiological agent is the key determining element in the epidemiology of the disease (Nurminen et al. 1999). The agents' concomitant with the infectious diseases can be categorized as living and non-living. The living agents comprises as viruses, bacteria, protozoa, rickettsia etc. Conversely, the non-living agents encompass heat, cold, water, nutrients, toxic substances etc. (Bethesda 2007). The infectious diseases are mostly ecological in origin as most of them are closely linked with the circumstances in the physical setting. It is the realities that environmental surroundings intensify the biological organisms' capability to thrive or feast. While some subsidiary situations occur in the natural environment however many are generated or improved by anthropogenic activities. Table 5.1 summarizes the known causes for a number of infections that have emerged recently and in the historical past.

Infectious diseases have continually been thoroughly interlaced with the antiquity of the developing countries which are idyllic locations for the emergence and proliferation of infectious diseases. Paucity, over-population, deforestation, urbanization, global warming and climate change, weak health structures are the physiognomies of any developing country and ideal locations for the appearance of infectious diseases.

Table 5.1 Recent example of emerging infections and probable factors in their emergence

Infection or agent	Factors contribution to emergence
Viral	
Argentine, Bolivian hemorrhagic fever	Changes in agriculture favouring rodent host
Bovine spongiform encephalopathy (cattle)	Changes in rendering processes
Dengu, denguhemorrhagic fever	Transportation, travel and migration; urbanization
Ebola, Marburg	Unknown
Hantaviruses	Ecological or environmental changes contact with rodent hosts
Hepatitis B, C	Transfusion, organ transplant, contaminated hypodermic apparatus, sexual transmission, vertical spread from infected mother to child
HIV	Migration to cities and travel; after introduction, sexual transmission, vertical spread from infected mother to child, contaminated hypodermic apparatus, transfusion, organ transplants
HTLV	Contaminated hypodermic apparatus, other

(continued)

Table 5.1 (continued)

Infection or agent	Factors contribution to emergence
Influenza (pandemic)	Possibly pig-duck agriculture, facilitating, reassortment of avian and mammalian influenza viruses
Lassa fever	Urbanization favouring rodent host, increasing exposure
Rift valley fever	Dam building, agriculture, irrigation, possibly change in virulence or pathogenicity of virus
Yellow fever	Conditions favouring mosquito vector
Bacterial	
Brazilian purpuric fever (Haemophilus influenzae)	Probably new strain
Cholera	In recent epidemic in South America, probably introduced from Asia by ship with spread facilitated by reduced water chlorination
Helicobacter pylori	Associated with gastric ulcers, possibly other gastrointestinal disease
Hemolytic Uremic syndrome	Mass food processing technology allowing contamination of meat
Legionella (Legionnaires disease)	Cooling and plumbing systems (Organism grows in biofilms that form on water storage tanks and in stagnant plumbing)
Lyme borreliosis (Borrelia burgdorferi)	Reforestation around homes and other conditions favouring tick vector and deer (a secondary reservoir host)
Streptococcus, group A (invasive; necrotizing)	Uncertain
Toxic shock syndrome	Ultra absorbency tampons
Parasitic	
Cryptosporidium	Contaminated surface water, faulty water purification
Malaria	Travel or migration
Schistosomiasis	Dam building
Leishmaniasis	Environmental factors, malnutrition, immunological, microbiological and pathological factors

Source Morse (1995)

5.4 Environmental Impacts of Controlling Disease Pattern and Distribution

The environment is the whole thing that ambiances the pathogen in its transmission from hosts to susceptible person or animal. The key environmental components comprising altitude, temperature, precipitation and relative humidity, vegetation conditions, influence the occurrence of growth, activity and durability of pathogens, vectors, zoonotic reservoirs of infection, and their interfaces with humans (Meade et al. 1988).

The landscapes involve of the larger physical configurations in the environment. These configurations are typically natural, but can be anthropogenic. Characteristics of the landscape that would affect disease spread most are the micro-climate, the occurrence of water and natures of vegetation. Landscape epidemiology encompasses the documentation of environmental areas where disease is transmitted. It is a universal method that comprises the interaction and connotations between components of the physical and cultural surroundings. This concept is first articulated by the Russian epidemiologists Pavlovsky (1966) suggested that the vegetation and geological circumstances essential for the conservation of specific pathogens in nature, one can use the landscape to identify the spatial and temporal distribution of disease risk. The following section provides some examples of emerging infectious diseases, considered under major categories of environmental and social influences, pertaining to climate, vegetation soil, land use and anthropogenic causes etc.

5.4.1 Climate

The climate and its intermittent deviations play a domineering role in disease diffusion. Climate is a major component of the environment and exerts a profound effect on vector-borne disease (Høye and Forchhammer 2008). It plays an important role in the seasonal pattern or temporal distribution of disease that are carried and transmitted through vectors because they often breed in particular climatic conditions (Elnaiem et al. 2003; Rahman 2008). Moreover, within their climatic limits, all the atmospheric features continuously disturb every facet of behavior, expansion and spreading of vectors. Normally, direct sunlight, an arid climate and high temperatures will lessen the subsistence times of pathogens in the environment.

- *Macro and micro climate and infectious disease*

When considering climate as a determining factor of disease a discrepancy is usually prepared between the macro-climate and the micro-climate. An enormous number of weather elements (temperature, rainfall, humidity, wind etc.) pool to create the macro-climate and can turn as a disease agent in their own right, either individually or in combinations. If the effects of weather on disease proxy and their transitional host or vectors are identified, it may be conceivable to envisage when host people are at a specific risk of constricting disease and thus to contrivance apposite control processes at premeditated times (McMichael 2015; Kim et al. 2007). The term micro-climate denotes the real climatic circumstances predominant in the exact, constrained environment where the host, agent, vector or intermediary hosts essentially living. However, macro-climate can have a direct influence on micro-climate, and is appropriate for the endurance of the disease agent and its vector or intermediate hosts.

Climate is the main factor influencing the spatial distribution, life cycle, seasonal activities, population dynamics and behavior of vector of any particular geographical area. Brown et al. (2007) found an inverse relation between increasing water temperature and Avian Influenza Virus (AIV) survival. Environments may not be apposite to spread year round, and several infections are seasonal, arising when the situation is promising to transmission. Mosquito-borne infections, like malaria, leishmanasis, and yellow fever are connected to rainy season (Bhunia et al. 2012a, b, c). The incidence of diarrheal diseases often rises with the first rains later the summer season. Ponds which fade in the summer season may in the moist season comprise water with snails that will transmit schistosomiasis (Moser et al. 2014). Though ecological factors such as geographical distribution adopt on whether hosts and pathogen co-occur, host immune capacity affects the triumph and sternness of the infection. Climate variables are identified to curb immune functions, and this has concerned for the creation, survival and imitation of the pathogen in hosts and vectors (Martin et al. 2010; Murdock et al. 2012).

- *Human-induced climate change*

Several pathogens and their vectors are very delicate to climatic conditions, mainly temperature, humidity and surface water bodies. It has become progressively certain not only that human face anthropogenic climate change, due to the persistent extreme emission of green house gases, but that the practice has instigated. Karl (2003) stated that the modern climate change is conquered by human impacts, which are now large enough to surpass the bounds of natural variability. The incidence and geographical range of particular plant and animal infectious diseases have apparently rehabilitated, at least partially in regards to climate change, over recent years (Harvell et al. 2002). For human infectious diseases, the underlying conformation is inherently more multifarious (involving many more demographic, social and technological effects), and therefore it has confirmed difficult to attribute clear-cut influences because of continued climate change. However, some reminiscent signal exists for an impact of recent climate change upon tick-borne encephalitis in Sweden, cholera in Bangladesh and malaria in parts of eastern Africa (Lindgren and Gustafson 2001; Patz et al. 2000; Rodo et al. 2002).

With the change of global climate, the conditions of life on Earth at the planetary scale have been changed. It will have sundry, mostly negative, concerns for biological systems universally and for dependent human societies.

5.4.2 Soils

Soils act indirectly as determinants of disease by causing starvation, if there is little or no vegetation or nutritional imbalances such as protein, energy, vitamin or mineral deficiencies. The distribution and abundance of vector seem to be influenced by structure and composition of soil along with its physical and chemical characteristics. Physical and chemical characteristics of the soil are leading the breeding and distribution of vector has been accentuated (Sivagnaname and Amalraj 1997). Some vectors require moist soil rich in organic and nitrogenous matter to breed

(Napier and Smith 1926; Adler and Theodor 1957). Subsequently, inorganic constituents of the soil were found to affect several breeding (Kesari et al. 1992).

5.4.3 Vegetation Condition

In the past, epidemiological studies often made use of information from local weather stations and some historical file based on observations. From space-borne satellite data, vegetation indices (VIs) were predominantly applied in studying the vector borne diseases. VIs develops the strong contrast in the reflectance of vegetation in the VIS and NIR wavelengths. These studies focus identification on priority regions for the rapid and precise study of the implication of vegetation over extensive areas as a means of comprehending and visualizing the current status of disease endemicity at present and near future projections. Such methods not only serve an indicator of the abundance of vectors (Malone et al. 2001; Wu et al. 2002; Odiit et al. 2006; Peterson et al. 2005; Raso 2006) but also point to other environmental factors, such as soil moisture regime, soil type, slope and elevation (Boone et al. 2000). An amalgamation of spectral indices of vegetation dynamics, surface reflectance, and estimations of surface temperatures have been employed by epidemiologists to figure out the vector ecosystems (Rogers et al. 2002).

The implication for landscape epidemiology is that, local to regional landscape information may be acquired frequently enough to study the seasonal changes in vegetation characteristics which may in turn be related to vector or host population dynamics. A further advantage of satellite data is the range of higher level data products that are derived from the data and made freely available to the scientific community. It is now possible for the users to download the radiometrically and geometrically corrected reflectance data and a range of derived products like vegetation index image, leaf area index image and so on. In this way researchers may choose to directly use these products in landscape epidemiology investigations without a detailed knowledge of the algorithms used to derive the products. For example, data from Moderate Resolution Imaging Spectroradiometer (MODIS) aboard NASA's Terra and Aqua satellites are helping scientists to routinely map the rate at which plant life on Earth absorbs carbon out of the atmosphere (http://neo.sci.gsfc.nasa.gov/).

Such maps effectively represent our planet's "carbon metabolism" called Net Primary Productivity (NPP). These maps show the location and amount of net carbon dioxide, which is the amount taken in by vegetation during photosynthesis minus the amount released during plant respiration. Such maps give a fascinating new insight into the intimate connection between the living world and the disease transmission area (Fig. 5.2). Because carbon dioxide is a key greenhouse gas that helps to warm our planet, climate scientists are concerned with drawing the flow paths of carbon dioxide. They are also interested in monitoring plant productivity as humans steadily increase the amount of carbon dioxide in Earth's atmosphere and cause global warming.

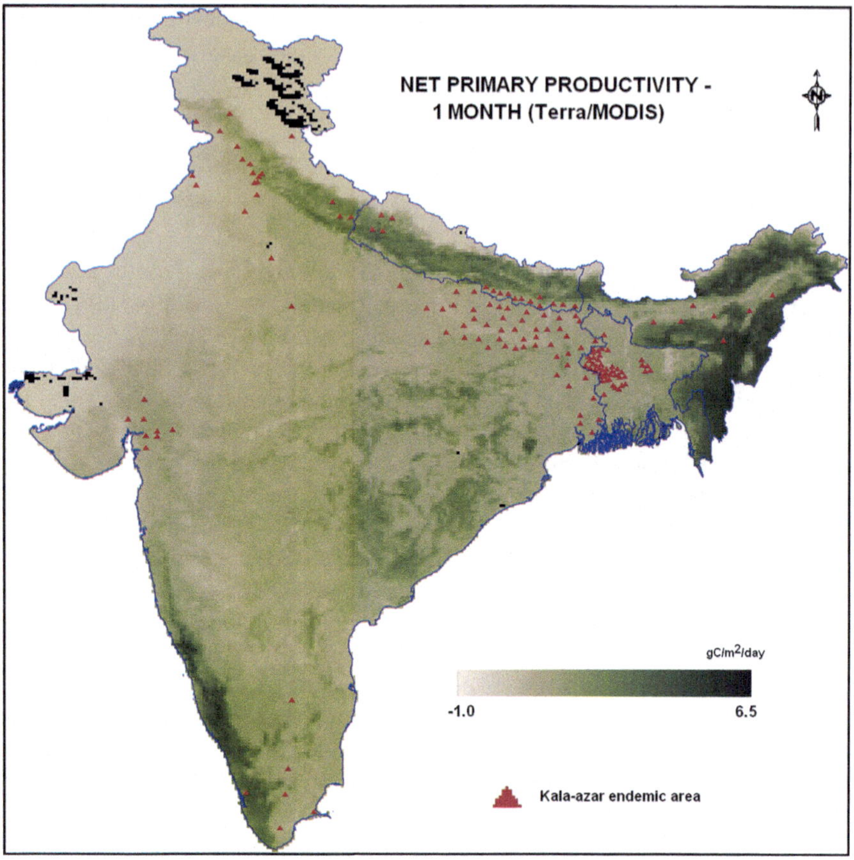

Fig. 5.2 Relation between visceral leishmaniasis (Kala-azar) endemic area and Net Primary Productivity, derived from MODIS/Terra (October, 2008) (*Source* http://modis.gsfc.nasa.gov/). The colors on these maps indicate how fast carbon was taken in for every square meter of land for the time span indicated. Values range from −1.0 g of carbon per square meter per day (tan) to 6.5 g per square meter per day (dark green). A negative value means decomposition or respiration overpowered carbon absorption; more carbon was released to the atmosphere than the plants took in. Black means "no data"

The result suggests that the present net primary production estimates in the KA endemic area are lower, and field with higher NPP have lower number of cases in South East Asia. On the other, the field with extremely less NPP has very rare cases of KA, and becomes eventually nil. For example, lower NPP indicates less photosynthetically active cover types; suggested as risk areas for Kala-azar endemicity (Connor et al. 1995).

Furthermore, NPP as the amount of Carbon fixed per unit area per unit time and is a useful proxy for measuring carbon sequestration by global ecosystems. The region has remarkably less NPP, indicates the higher concentrations of carbon monoxide, influence the temperature locally. The impact of temperature on sand fly populations is rapid and direct. Using NPP, scientists were able to predict the distribution vegetation condition and synoptic temperature with an overall accuracy of more than 80% (Thanyapraneedkul et al. 2009). The scenario of the NPP changes throughout the region resulted in the change of photosynthetic radiation and temperature could be helpful to predict the vector distribution. As a result, we also conclude that remotely sensed data is useful in providing a robust and quick estimate of vegetation condition, ground carbon stocks, and temperature in terrestrial ecosystems.

5.4.4 Biodiversity Change and Habitat Fragmentation

Deforestation, with destruction of habitat, increases the 'edge effect', which then endorses pathogen–vector–host contact. In the recent decades, this progression has contributed to the emergence of the several viral haemorrhagic fevers in South America (McMichael 2004). The influence of forest clearance, with road construction, ditch creation, and succeeding stemming and irrigation, is known to have various influences on anopheline mosquito species. Empty land and the formation of ditches may enrich breeding prospects for the foregoing indigenous malaria-transmitting anopheline mosquitoes. By contrast, habitat demolition may eradicate certain local mosquito species, possibly thereby opening a niche for an invasive anopheline species (Povoa et al. 2001).

5.4.5 Niche Invasion

The appearance of certain infectious diseases results from a pathogen conquering a different or newly emptied niche. A decent instance is the Nipah virus, which appeared as a human disease in Malaysia in 1999, instigating over 100 deaths (Chua et al. 2000). This highly pathogenic virus appeared from its natural reservoir host species (fruit bats) via in land animal (pig) amplifier hosts. The ecological trigger seems to have been multifaceted sequences of human amendments to fruit bat habitat and agriculture in amalgamation with a period of drought (Daszak et al. 2001; Chua et al. 2002).

5.4.6 Ecosystem Modifications, Loss of Predators and Host Species Imbalance

Forest fragmentation has directed to vicissitudes in biodiversity. Lyme disease exemplifies this kind of factor. The ticks usually feed on deer and white-footed mice, with the latter being the more proficient viral host species. This comprises the loss of various predator species—foxes, wolves, raptors and others—and a subsequent alteration of ticks from the less to the more proficient host species. These variations, along with middle-class suburban sprawling to woodlands, have all been interrelated in the incidence of this disease (Schmidt and Ostfeld 2001).

5.4.7 Anthropogenic Causes

Landscapes and human circumstances are closely connected. The human environments are generated by an amalgamation of natural components and how people have amended these components. Many contaminations are interconnected to particular environment, and people with definite livelihoods, socio-economic status, femininity, or faith may be far more at danger than others.

People acclimatize their environments to their necessities. If these variations are well done, they can support to thwart the diffusion of disease. In practice they often reassure the transmission of disease, conversely, as people do not have the space for inspiration, indulgent, time energy, or economic or material worth to ensure them appropriately.

5.4.8 Host Transfer

This is the ancient story of pathogens 'jumping ship'. The HIV/AIDS pandemic has prompted us of this continuing risk. Bush meat stalking in Africa has directed to extra confined emergence occurrences e.g., forest labor scutting up chimpanzee meat have become infected with Ebola virus (Patz and Wolfe 2002; WHO 1996).

Cross-species transmission is bidirectional. For instance, the parasitic disease, Giardia was known to Ugandan mountain gorilla by humans through ecotourism and conservation activities (Nizeyi et al. 1999). Non-human primates have assimilated measles from ecotourists (Wallis and Lee 1999).

5.4.9 Land Use and Environmental Change

Landscape variables such as land use/land cover (LULC) may be mainly imperative since they are openly interconnected to vector presence (Feliciangeli et al. 2006). The growing scale of our interference in the environment, both intentionally (e.g. land defrayal, urbanization) and as collateral influence (e.g. global climate change, species extinctions), is unsurprisingly hastening the rate of appearance of new infectious diseases. The main human-induced environmental vicissitudes that disturb infectious disease risk contain: tropical deforestation; road construction; irrigation; dam construction; local/regional weather irregularities; increased crop and animal production schemes; urban sprawl; unremitting poor sanitation; and contamination of coastal zones (Patz and Confalonieri 2005). Bhunia et al. (2012a, b, c) investigating the role of land use/land cover (LULC) for leishmanias is transmission and their suitability for vector habitats resulting in a framework highlighting the links between LULC and areas endemic for the disease at different scales.

5.4.10 Rehabilitated Habitat, with Propagation of Reservoir or Vector Populations

Worldwide change includes increasing numbers of human immigrants and travellers, augmented world trade, progresses in global communication and transportation affects the scale and tempo of change in health risk concerning to invasive viruses and diseases (McMichael and Bouma 2000). Anthropogenic actions can generate appropriate new habitats within the normal range of the vectors. An amalgamation of the novel and pre-existing habitats can then pose to growth in the vector population and, hence, disease transmission (Ramasamy and Surendran 2016). Growing disturbance in urban and exurban environments enables outlines and fast spread of new invasive species, ensuing in habitat fragmentation and abridged biological diversity (Hansen et al. 2005). Alteration to conquer newly designed anthropogenic ecological niches has been proposed to be connected with current speciation in major African vectors of the *Anopheles gambiae* complex (Kamdem et al. 2012). Conversely in Burkino Faso, biogeographical aspects such as humidity rather than anthropogenic changes seem tohave prejudiced the variances in ecological niches employed by the similar two species (Costantini et al. 2009).

5.5 Case Study 1: Correlative Analysis of Geo-environmental Factors and *Phlebotomus argentipes* Distribution of Vaishali District

5.5.1 Introduction

The geographical distribution of sand fly (*Phlebotomus argentipes*), their life history, host and ability to transmit the infections are determined by both intrinsic and extrinsic factors (Lewis 1971). Previously, it was recommended that the distribution of vector *P. argentipes* in Indian subcontinent is particularly sensitive to climate, non-climatic factors and demographic characteristics (Bhunia et al. 2010a, b; Picado et al. 2010a, b, c; Kesari et al. 2010; Ranjan et al. 2005). Sand fly reproduces best between temperatures 25 and 30 °C and relative humidity 80–85% (Picado et al. 2010a, b, c; Singh et al. 2008a, b). During the warmer months the density is less (Napier 1926; Ranjan et al. 2005), as the temperature in the area ranges between 40 and 46 °C; and the species also disappeared during the winter months (Smith 1959). However, the environmental variables are considered to be as dynamic, as it is change for a particular period of time. Though the variables considered in our analysis are static, however, it has a significant prospective based on the seasonal pattern of features distribution, until the major natural hazard has occurred in this area. Conversely, due to the lack of sufficient data at shorter temporal intervals seasonal summaries were chosen based on the sand fly abundance (e.g., peak/high and lean/low) to analyze their habitat characteristics in relation to environmental factors in this particular VL transmission region.

Many remotely sensed abiotic and biotic environmental variables are relevant for the study of Kala-azar transmission and habitat niches of the vector (Nieto et al. 2006; Colacicco-Mayhugh 2010). Earlier authors suggested that the presence of vegetation is a risk factor for Kala-azar (Dinesh and Dhiman 1991; Dhiman and Dinesh 1992). Peri-domestic vegetation thus constitutes a significant risk since the trees provide shade, which not only creates dark and humid atmosphere around the vicinity of the houses, but also suitable resting sites for sand flies. Furthermore, plants rich in fructose attract *P. argentipes* and it is not unusual that fructose-containing climber plants are propped up by bamboo, which is often grown in these surroundings (Dhiman and Dinesh 1992; Ranjan et al. 2005; Sudhakar et al. 2006).

The wider use of Remote sensing (RS) and Geographic Information Systems (GIS) is needed to better understand the potential distribution of the vector in an endemic Kala-azar focus and too rapidly, precisely and safely direct operations to these areas (Elnaiem et al. 2003; Gebre-Michael 2004). RS data have been used as an adjunct in various epidemiological studies (Herbreteau et al. 2005; Kalluri et al. 2007) but has so far not been fully utilized with respect to Kala-azar. Nevertheless, ecological indicators have been exposed by this approach (Victora et al. 1997; Sudhakar et al. 2006, b), revealing the role of environmental markers such as vegetation vigor, soil characteristics, humidity, etc. (Elnaiem et al. 2003; Singh et al. 2008a, b).

Present study investigated the correlative analysis of geo-environmental factors in relation to *P. argentipes* abundance in VL endemic district of Bihar (India).

5.5.2 *Materials and Method*

5.5.2.1 Study Area

The Vaishali district is spread over 2,036 km^2 area and is located between 25° 41′ N–25.68 °N latitude and 85° 13′ E–85.22 °E longitude (Fig. 5.3). The climate of the district is sub-tropical to sub-humid. The district generally enjoys a bracing and healthy climate with three well marked seasons—winter, summer and the rainy season. Humidity increases and the rains set towards the middle of the June. It ranges between 60 and 90%. The rainy season lasts till the end of September or the middle of October. The average annual rainfall in the district is 1168 mm. The soil is most fertile and suitable for cultivation of high yielding crops like sugarcane and

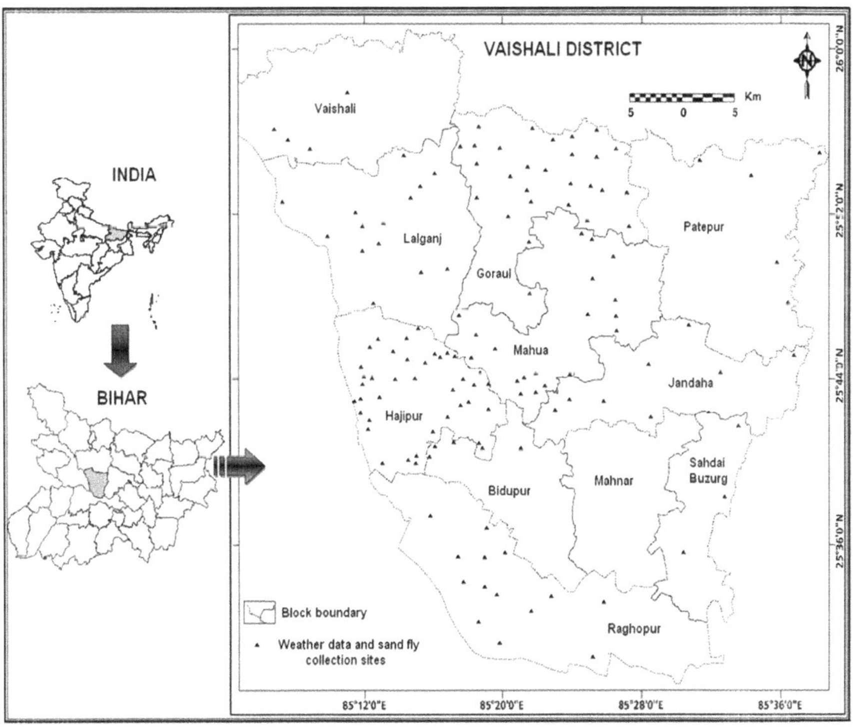

Fig. 5.3 Map of the study area

wheat. Quaternary Alluvial deposit consisting of alternate layers of sand, silt, clay and gravel forms prolific unconfined and confined aquifer system. The district is basically agricultural in character. Rice, maize and wheat are the main crops of the district.

5.5.2.2 Vector Density Estimation

The adult sand flies were collected from January 2008 to December 2011 in the study area. The sand fly collection was made for two seasons (e.g., lean/winter and peak/post-monsoon) separately for the same villages, and in similar collection sites. In order to determine sand fly density, flies were collected through CDC light traps from two indoors (living room and cattle shed) resting place. Collected sand flies were stored into 70% ethanol in vials which were labeled for the area, village name, and number of sand flies caught. All species were mounted on a micro slide using Canada balsam, as a measuring media (Remaudière 1992). For species identification Lewis (1978) was followed. The average density was calculated through the total number of sand fly collection for the selected site divided by the total number of traps installed × number of day collection for the respective sites/villages for each season.

5.5.2.3 Climatological Data Source

The climatic data consisted of indoor climate (e.g., room temperature and relative humidity) for winter (i.e., lean) and post-monsoon (i.e., peak) seasons, which were obtained from 139 villages of Vaishali district (Fig. 5.3) using Polymeter (BARIGO, Model No. 305, Made in Germany) from the ground at the time of sand fly collection. For each village, climate data were recorded from the ten houses where the traps were installed and then averaged it, to represent the local climate of the respective village.

5.5.2.4 Spatial Statistical Analysis

For indoor climatic data (temperature and relative humidity), local polynomial interpolation using all points only within the defined neighbourhood was used based on the weighted value (e.g., recorded from the field). However, in this interpolation technique short-range of variation within the dataset has been removed. The root mean square error also calculated to represent the predict error e.g. the difference between the prediction and the actual difference value.

5.5.2.5 Satellite Data Acquisition and Analysis

For the present study, Landsat5 satellite data was used which is near-polar, sun-synchronous orbit and derived from the USGS Earth Explorer community (http://earthexplorer.usgs.gov/). Landsat5 satellite carries Thematic Mapper (TM) sensor that provides spatial resolution in six spectral bands ranging from blue to the middle infrared, and a 120 m resolution thermal infrared band (Markham et al. 2004). Since the imagery (Path/Row 141/42) for February (Date of Pass—12-02-2010) and October (Date of Pass—22-10-2009) were selected for lean and peak season analysis respectively. The satellite data were geometrically corrected to the Universal Transverse Mercator (UTM) projection, Zone 45 and the World Geodetic Survey (WGS) 84 datum. The root mean-square error (RMSE) was about 0.3 pixels which is suitable according to Lunetta and Elvidge (1998). All images were subsets and masked to demonstrate the Vaishali and Muzaffarpur district only. The whole image processing operations were accomplished using ERDAS IMAGINE 9.2 (Leica Geosystems 2008) software. The study methodology is represented in Fig. 5.4.

5.5.2.6 Geo-environmental Parameters Characterization and *P. argentipes* Habitat Suitability

The geo-environmental indices, like indoor climate (e.g., temperature and humidity), RDVI (Roujean and Breon 1995), LST (Sobrino et al. 2004; Artis and Carnahan 1982), WI (Crist and Cicone (1984) was calculated to estimate the vegetation vigor, land surface temperature, surface dampness respectively. To find out the relationship between the geo-environmental indices of the seasons and the spatial distribution of *P. argentipes* density, a 500 m buffer zone was generated for each survey point. The association of *P. argentipes* density with the minimum, maximum, and mean values of each geo-environmental index was calculated. Alternatively, NDPI (Lacaux et al. 2007) was used to extract the surface waterbody within the study site. Location-allocation map of seasonal water body was prepared by extracting the inland waterbodies, like perennial river, seasonal flowing water courses, areas occupied with stagnant water (i.e., low laying areas) and other moist areas (i.e., paleochannels, marshy land etc.). Small waterbodies, such as ponds were not considered in this theme. The location-allocation analysis was performed to measure the efficiency of the inland surface water bodies. The efficiency was calculated with a maximum of 1 km distance from each spatial feature to determine the suitable sites of disease location in relation to inland water bodies. Land use/land cover (LULC) maps were produced to identify the different classes of land from Landsat TM imagery based on supervised classification technique with Maximum Likelihood (MXL) Algorithm (Belward and de Hoyos 1987; Richards 1986). To correctly perform classification accuracy assessment, an error matrix table was derived, and based on the error matrix table classification accuracy was determined (Congalton 1991; Rosenfield and Fitzpatrick 1986). All image processing was

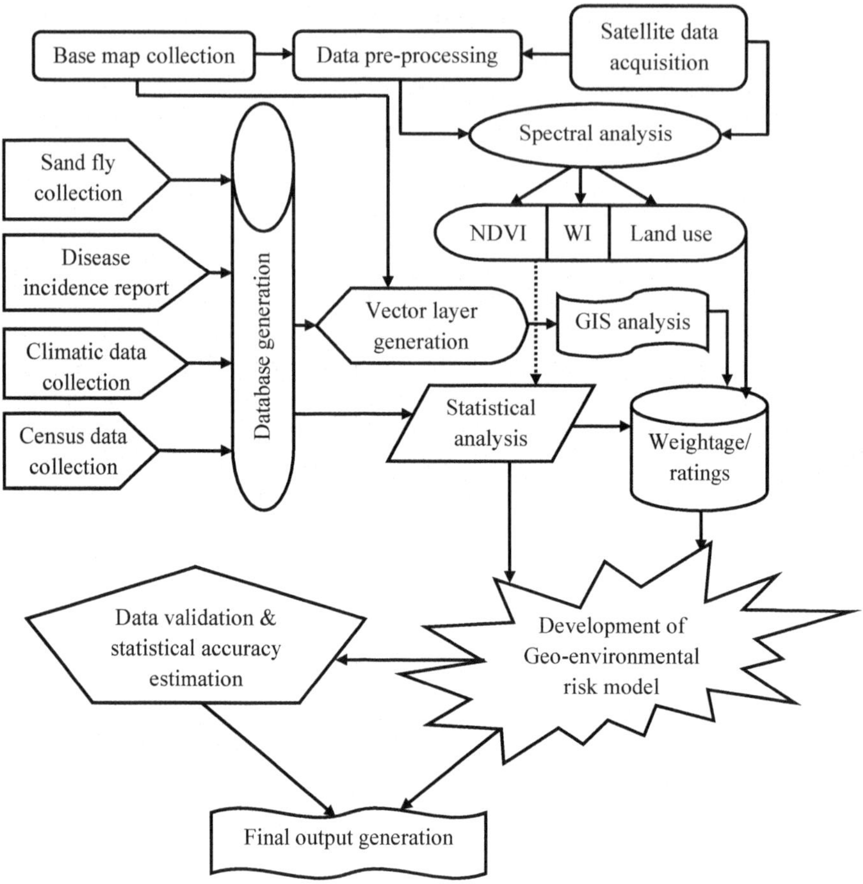

Fig. 5.4 Flow chart map of the study methodology

completed using ERDAS IMAGINE (Leica Geosystems 2008) software v9.2. The LULC classes were ranked according to their *P. argentipes* density habitat suitability, e.g., indicating their potential risk for Kala-azar transmission based on the class density. The class density is the number of times presence of classes within 500 m buffer zone divided by the total number of such times considered for calculation (i.e., divided by 140 numbers). A univariate analysis (Pearson correlation co-efficient) and linear regression analysis was performed to understand the statistical significant association between *P. argentipes* abundance and geo-environmental variables. The statistical significance was defined as $P < 0.05$.

5.5.2.7 Static-Prescriptive Model for Demarcating Sand Fly Abundance Area

The model deals with the spatial data in a particular period of time and offers a prediction of sand fly abundance. To differentiate sand fly abundance zone, all these layers were integrated using weighted index overlay method (Bhunia et al. 2012a, b, c; Hongoh et al. 2011; Tsiko and Haile 2011; Ying et al. 2007; Rakotomanana et al. 2007). Considering the influence on the sand fly abundance, the weights of different themes were allotted on a scale of 1–5. Different classes of each theme were assigned weights on a scale of 1–5 according to their relative influence on *P. argentipes* abundance. Using this scale, a qualitative assessment of different classes of a given theme was performed, with: very low abundance (weight = <1.0); low abundance (weight = 1.5–2.0); moderate abundance (weight = 2.5–3.0); high abundance (weight = 3.5–4.0); and very high abundance (weight = >4.5). Subsequently, a pairwise comparison matrix was built *via* Saaty analytical hierarchy process (AHP 1980) to calculate normalized weights for individual themes and their features. In the present study, the hierarchy model was anticipated to integrate information from geo-environmental and climatic factors using Saaty AHP, and offer location of probable sand fly abundance sites. Furthermore, the conveying weights using pairwise assessment were more appropriate than the straight obligation of the weights, because one can verify the reliability of the weights by scheming the consistency ratio. The consistency ratio was calculated by the ratio of the index of coherence in the initial matrix for the random index in the matrix with the same dimension, proposed by Saaty (1980).

The total weights of different polygons in the integrated layer were derived from the following equation to obtain the potential sandfly abundance index:

$$SAM = (IRT_w * IRT_{wi}) + (IRH_w * IRH_{wi}) + (LST_w * LST_{wi}) + (RDVI_w * RDVI_{wi}) + (WI_w * WI_{wi}) + (LULC_w * LULC_{wi}) + (LASWB_w * LASWB_{wi})$$

where, SAM = sand flies abundance mapping, IRT = inside room temperature, IRH = inside room humidity, LST = land surface temperature, RDVI = re-normalized difference vegetation index, WI = wetness index, LULC = land use/ land cover, and LASWB = location-allocation analysis of seasonal waterbody, and the subscripts '*w*' and '*wi*' refer to the normalized weight of a theme and the normalized weight of the individual features of a theme, respectively. The range of SAM values was divided into three equal classes (called zones) and the SAM of different polygon features under different range was grouped into one class.

5.5.2.8 Model Validation and Comparison

The final risk model for peak and lean season in Vaishali district was summarized for delineating probable 'high' and 'low' *P. argentipes* abundance area based on the geometric interval of calculated values. The calculated index value was categorized into five divisions for easier interpretation of results and model validation. The model was validated through the collection of *P. argentipes*in the lean (January and February) and peak season (September and October) for the years of 2012 in the study area. *P. argentipes* density of the Vaishali district was collected using CDC light trap from the 25 villages for each peak and lean season separately. In the Muzaffarpur district, *P. argentipes* densities were recorded from 30 to 51 villages from the lean and peak season respectively. A point layer was generated based on the collected sand fly zone abundance in the study area on GIS platform. The collected sand fly abundance was compared with the lowest abundance sand fly zone versus high sand fly abundance zone, and an error matrix table was generated. The producer accuracy and user accuracy for each identified class were calculated and the overall accuracy was estimated.

5.5.3 *Results*

5.5.3.1 *P. argentipes* Density in Lean and Peak Season

A total of 1,169 sand flies those belonging to the genus *Phlebotomine* and *Sergentomyia* were collected (Table 5.2). Among the three species of the study site 62.39% were *P. argentipes* (male: female ratio 1:1.35), 37.05% was *Sergentomyia* (male: female ratio 1:1.86) and 0.56% was *P. papatasi* (male: female ratio 1:0.84). During this study period, the aggregate population of sand flies was found to be lowest in December to February month (i.e. lean season). The sand fly population rose during March, with highest peaks of sand fly density were observed between August and September (i.e., peak season) in the Vaishali district. In lean season, the average sand fly density of the Vaishali district was 0.71 (95% CI: 0.02–2.00) with a standard

Table 5.2 Month wise collection of *Phlebotomus argentipes* in Vaishali district, Bihar, India

Month	Total number of *P. argentipes* captured	% of Male *P. argentipes*	% of Female *P. argentipes*	Percent total
January	12	0.88	0.63	1.52
February	11	1.14	0.25	1.39
March	95	4.04	7.95	11.99
April	55	3.54	3.41	6.94
May	49	4.04	2.15	6.19
June	75	3.91	5.56	9.47
July	85	2.78	7.95	10.73
August	62	3.66	4.17	7.83
September	103	5.93	7.07	13.01
October	144	8.46	9.72	18.18
November	87	3.16	7.83	10.98
December	14	1.01	0.76	1.77
Total	792	42.55	57.45	

deviation of ±0.56. In peak season, highest sand fly density (10.85 per trap/night) was recorded from the SenduariGobind village of Mahua PHC and the lowest sand fly density (0.45 per trap/night) was recorded from the ChakSalaeh village of Hajipur PHC. Conversely, the lowest sand fly density was recorded in Kajribhat village of Jandahar PHC (0.05 per trap/night); while the high density (2.00 per trap/night) was documented from the Bahrampur village (Raghopur PHC) in the lean season.

5.5.3.2 Seasonal Climatic Variables

- *Lean Season*

IRH of Vaishali district of the lean season varied from 34.5 to 54.6%, with a mean IRH of 45.00% (SD ± 4.76). The district has been categorized into 4 divisions based on geometric interval, such as (i) <40%, (ii) 40–45%, (iii) 45–50%, and (iv) >50% (Fig. 5.5). Within the study area, the maximum relative humidity was recorded from the western and north-central part, and some small pockets were also observed in the northeastern part. Less IRH (e.g. less than 40%) was recorded from the southeastern part of the district. The correlation between IRH and sand fly abundance in lean season of Vaishali district showed a positive and significant relationship ($R^2 = 0.45$; $P < 0.05$).

RT of lean month within the Vaishali district ranged from 15.20 to 25.50 °C, with an average temperature of 21.50 °C (SD ± 1.50). Based on the distribution of IRT, the entire district has been divided into 4 categories, i.e. (i) <19.50 °C, (ii) 19.50–22.00 °C, (iii) 22.00–24.50 °C, and (iv) >24.05 °C (Fig. 5.5). However, the maximum IRT in lean season was recorded from the western, southern and

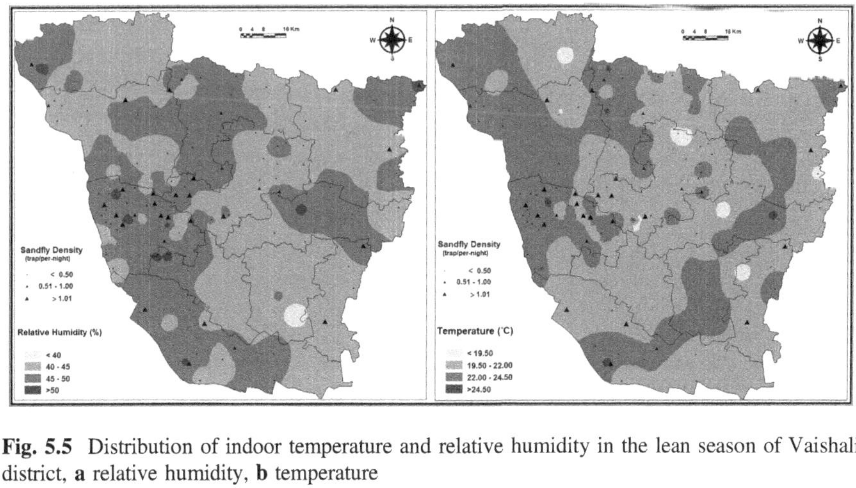

Fig. 5.5 Distribution of indoor temperature and relative humidity in the lean season of Vaishali district, **a** relative humidity, **b** temperature

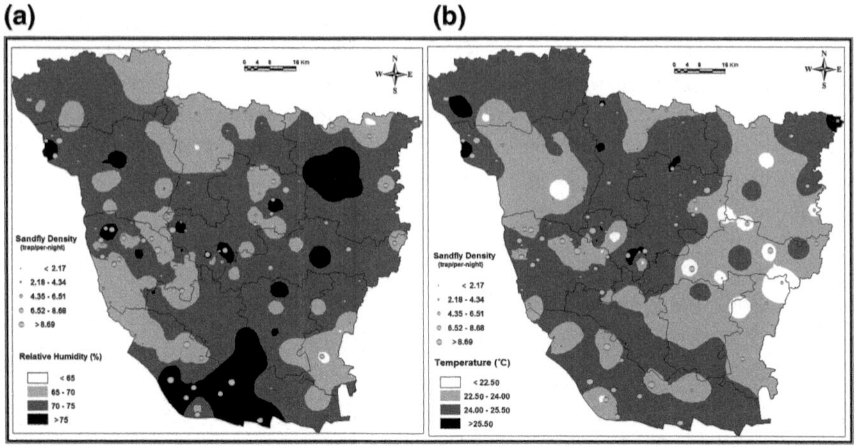

Fig. 5.6 Distribution of indoor temperature and relative humidity in the peak season of Vaishali district, **a** relative humidity, **b** temperature

central-east of the district. Minimum IRT (e.g., <19.50 °C) was recorded from the very small pockets of the central Vaishali district. Moreover, IRT of the lean season showed a statistically significant relationship with the abundance of vector, *P. argentipes* density ($R^2 = 0.23$; $P < 0.05$).

- *Peak season*

In the peak season, the IRT of Vaishali district is varied from 21.00 to 26.80 °C (Fig. 5.6). The average IRT of the entire district was 24.57 °C (S.D. ± 1.46). The small pockets of maximum IRT were recorded from the northwestern, northeastern and central part of the district. However, highest sand fly density was recorded from the district contained temperature of 24.00–25.50 °C. Furthermore, the minimum temperature was recorded from some small pockets of the eastern part of the district. IRT of peak season in Vaishali district illustrated a significant association with the sand fly density ($R^2 = 0.37$; $P < 0.05$).

In the Vaishali district, IRH is ranged from 62 to 86% in the peak season (Fig. 5.6). The mean IRH of the district is recorded as 71.32% with a standard deviation of ±5.88. Based on the IRH value, the entire district has been categorized into four parts, like (i) less than 65%, (ii) 65–70%, (iii) 70–75%, and (iv) more than 75%. However, the highest IRH was documented from the southern and north-eastern part of the district. IRH of peak season have a significantly ($R^2 = 0.59$; $P < 0.05$) influence on the *P. argentipes* density distribution in the study site.

5.5.3.3 Vegetation Vigor and Its Relation to *P. argentipes Abundance*

The value of RDVI in the lean season has ranged from 0.08 to 1.17, with an average value of 0.11 (SD ± 0.03). The value of RDVI in the peak season is varied from

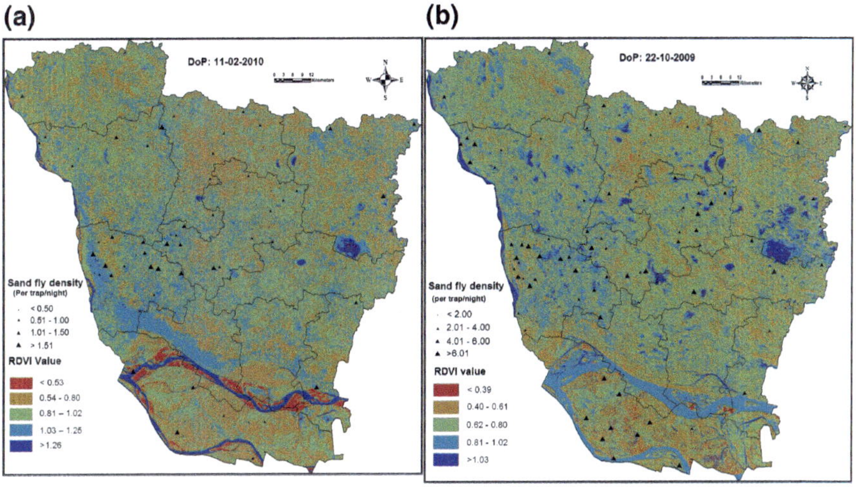

Fig. 5.7 Estimation of renormalized difference vegetation Index (RDVI) in relation to vector abundance of Vaishali district, **a** lean season, **b** peak season

0.07 to 1.15 (mean ± standard deviation 0.10 ± 0.02). However, the higher value of RDVI indicated the vegetation density of this region is high and vice versa. In the peak season, the higher RDVI value was recorded in the southwestern part of the district, while the lowest value was evidenced in the southern part of the district. Moreover, the lowest RDVI value was also found in the northern and northeastern part of the district. As shown in Fig. 5.7, the Pearson's correlation coefficient analysis between the spatial distribution of *P. argentipes* density and RDVI composite values in the peak season had a significant negative association with the maximum and mean RDVI value ($r = -0.530$; $P < 0.002$; $r = -0.660$; $P < 0.000$). On the other hand, a very strong negative correlation was observed between mean RDVI values and the spatial abundance of *P. argentipes* ($r = -0.549$; $P < 0.005$) in the peak season. Similarly, information was available from the RDVI values of peak season in which the maximum RDVI values were negatively associated with *P. argentipes* abundance ($r = -0.49$; $P < 0.0036$). The minimum RDVI of peak season, on the other hand, had a negative correlation with the spatial distribution of *P. argentipes* density in the peak season ($r = -0.52$; $P < 0.001$), while marginal association was observed in the lean season ($r = -0.16$; $P < 0.05$).

5.5.3.4 Land Surface Temperature (LST) and Its Relation to *P. argentipes* Abundance

Figure 5.8 shows the spatial distribution of LST derived from Landsat—5 TM at different season of Vaishali district. The LST ranged from 19.00 to 34.00 °C (mean ± standard deviation—26.50 °C ± 4.76) for peak season and 19.00–31.00 °C

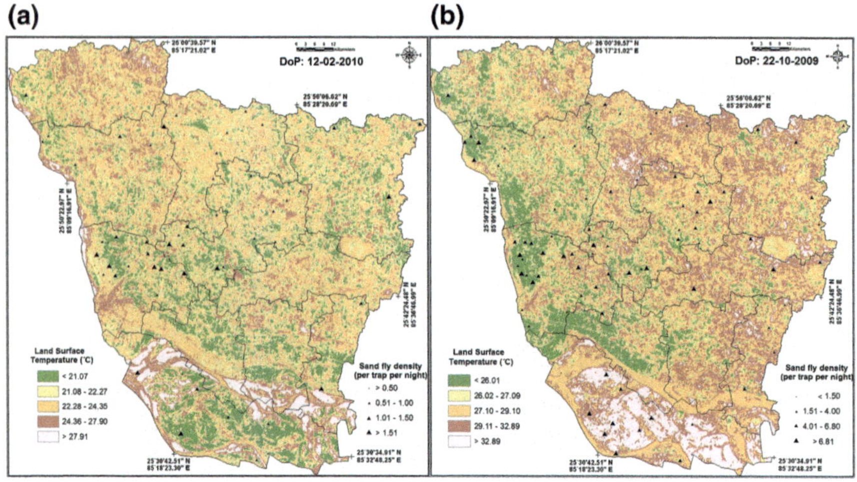

Fig. 5.8 Estimation of LST in relation to vector abundance of Vaishali district at different seasons, **a** lean season, **b** peak season

(mean ± standard deviation 25.00 ± 3.89) for the lean season. Results showed a very strong and positive relationship between minimum and mean LST values, e.g., $r = 0.68$ and $r = 0.59$ respectively in peak season. Moreover, maximum LST values showed moderate and insignificant relationship with the *P. argentipes* abundance ($r = 0.35$; $p < 0.08$). Alternatively, in lean season, strong and positive correlation also existed between *P. argentipes* abundance and maximum LST ($r = 0.58$, $p < 0.006$), followed by mean LST values ($r = 0.52$, $p < 0.020$). Overlaying the LST map on the spatial distribution of *P. argentipes* density demonstrated that areas with LST values of 22.27–24.35 °C, generally coincided with areas with high numbers of *P. argentipes* density (Fig. 5.5) in lean season. Alternatively, in peak season, maximum sand flies density of *P. argentipes* was recorded with LST values 27.09–29.11 °C.

5.5.3.5 Wetness Index (WI) and Its Relation to *P. argentipes* Abundance

Wetness index map was prepared to investigate the surface dampness of study area (Fig. 5.9). The WI values of the study area are varied from −97.79 to 48.92 (mean ± standard deviation −11.63 ± 34.59) in peak season. In the lean season, the value of WI is ranged from −91.45 to 33.51 (mean ± standard deviation −16.03 ± 29.13). The minimum value of WI indicates the dryness (e.g. sandy area), whereas, the maximum value indicates wetness (e.g. water body/river) within the study area. Maximum, minimum and mean WI index value was calculated for each buffer zone. The Pearson correlation test showed a significant negative

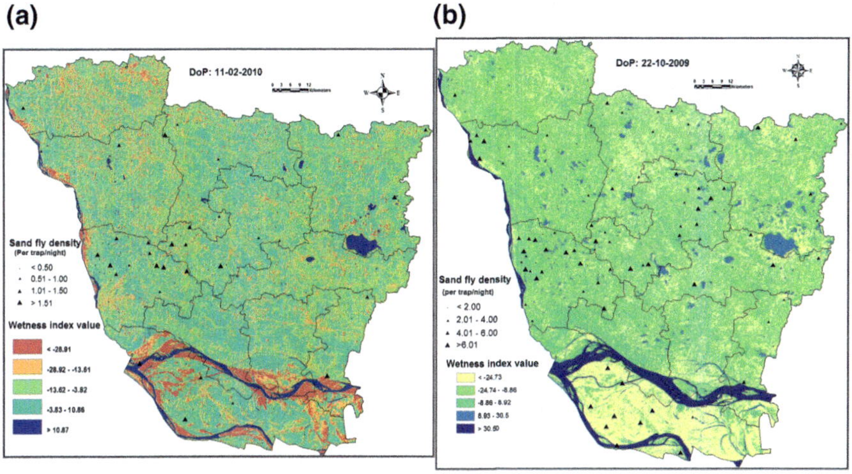

Fig. 5.9 Estimation of wetness index (WI) in relation to vector abundance of Vaishali district, **a** lean season, **b** peak season

relationship with and maximum and mean WI value with the *P. argentipes* abundance ($r = 0.64$ and 0.37 respectively) in the peak season. However, no significant relation was found with the minimum WI value ($r = 0.17$). Alternatively, in the lean season, maximum ($r = -0.54$) and mean ($r = -0.40$) WI value showed strong and negative association with the *P. argentipes* abundance; while, meager association was found with the minimum value of WI ($r = -0.04$). Overlaying the WI map on the spatial distribution of *P. argentipes* abundance established that areas with WI values of −8.86 to 8.92, usually corresponded with areas with high numbers of *P. argentipes* density (Fig. 5.6b) in peak season. On the other hand, in lean season, maximum sand flies density of *P. argentipes* abundance was evidenced from WI values of −11.00 to 5.77 (Fig. 5.9).

5.5.3.6 Land Use/Land Cover (LULC) and Its Relation to *P. argentipes* Abundance

The LULC map of the study site is presented in Fig. 5.10. Following land use/land cover classes were considered in image classification: river, sand, settlement, permanent/surface water body, marshy land, moist fallow, dry fallow, crop land, agricultural fallow, vegetation, plantation with settlement, and dense forest (Table 5.3). In the Vaishali district, the crop and agricultural land are covered by 65.03 and 57.95% area in the lean and peak season respectively; and the remaining area covered by vegetation, moist fallow, river, plantation with settlement, dry fallow, surface water body and marshy land. An accuracy assessment of the results, using additionally known land cover sites as test areas (5 test polygon per class), illustrated an overall classification accuracy of 85.82% in lean season, while in the

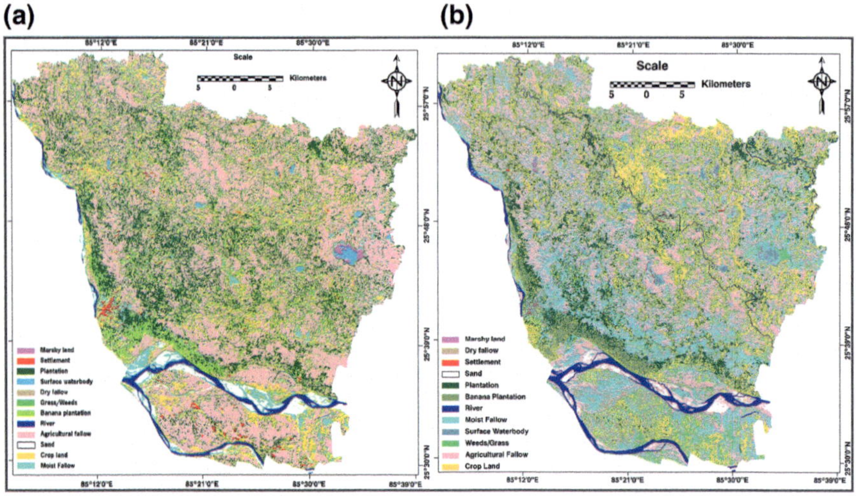

Fig. 5.10 Land use/land cover map of Vaishali district, **a** lean season, **b** peak season

peak season it was 87.70%. Kappa statistics were estimated for each land cover category in each season. The value of Kappa was 0.84 in lean season; while, in the peak season it was 0.86.

Twelve LULC classes, with special emphasis on crops, vegetation and fallows, were investigated for suitability with respect to *P. argentipes* abundance with the overall aim to score them for endemic risk in the lean season. In the peak season, marshy land and settlement recorded the highest class density.

5.5.3.7 Location-Allocation Analysis of Seasonal Water Body and Its Relation to *P. argentipes* Abundance

A location-allocation map of Vaishali district was extracted from the NDPI image for the peak and lean season separately (Fig. 5.10). In the Vaishali district, the perennial river is located on the southern (i.e., the Ganges River) and western part (i.e., Gandak River) of the district. A small perennial river (i.e., Burhi Gandak) was found to be flowed on the central part of the district. Furthermore, the region contains many seasonal flowing water courses, flowing from the northeast corner towards the southeast corner of the district. Some of the seasonal flowing water courses were also observed on the north-west corner of the district. The large and small low laying areas (locally known as '*chaurs*') were found across the entire district and the other moist areas (i.e., peleochannels) were found on the both sides of the river courses that contained water for the whole year. During the lean season, inland water bodies of Vaishali district and their areal coverage was calculated as

Table 5.3 Land use/land cover features of Vaishali district (computed from Landsat5 TM data)

Land use/land cover classes	Area (in Hect.)	% of area covered	Class density (%)	Suitability
Lean season				
River	3219.84	1.58	0.22	1
Sand	4358.52	2.14	1.40	1
Surface waterbody	1265.13	0.62	2.04	1
Marshy land	412.29	0.20	0.10	1
Plantation/sparse vegetation	25100.91	12.30	8.55	3
Banana plantation	26751.69	13.11	9.23	3
Grass/weeds	11308.68	5.54	18.82	4
Dry fallow	2155.50	1.06	11.55	3
Moist fallow	9574.65	4.69	24.35	4
Settlement	1014.66	0.50	4.01	2
Agricultural fallow	96282.54	47.18	3.57	2
Crop land	22616.19	11.08	16.16	4
Peak season				
River	5669.37	2.78	0.62	1
Sand	1601.91	0.78	1.97	1
Surface waterbody	782.28	0.38	11.99	3
Marshy land	255.33	0.13	14.76	4
Plantation/sparse vegetation	18685.26	9.16	6.11	2
Banana plantation	3324.87	1.63	4.68	2
Grass/weeds	17683.29	8.67	9.95	3
Dry fallow	9025.47	4.42	12.30	3
Moist fallow	44259.3	21.69	3.93	2
Settlement	482.94	0.24	27.18	4
Agricultural fallow	55113.39	27.01	2.99	2
Crop land	47182.14	23.12	3.54	2

follows: (i) perennial river (49.19 km^2), (ii) seasonally flowing water courses (10.52 km^2), (iii) areas occupied with stagnant water bodies (33.59 km^2) and (iv) other moist areas (3.42 km^2). In the peak season, inland water bodies of Vaishali district were also extracted from the NDPI image and the area calculated for each water bodies as follows, (i) perennial river (100.97 km^2), (ii) seasonally flowing water courses (16.72 km^2), (iii) areas occupied by stagnant water bodies (73.42 km^2) and (iv) other moist areas (10.29 km^2). Overlaying the location-allocation map onto that depicting the spatial distribution of *P. argentipes* abundance confirmed that areas with seasonally flowing water courses generally coincided with areas with high *P. argentipes* density (Fig. 5.11).

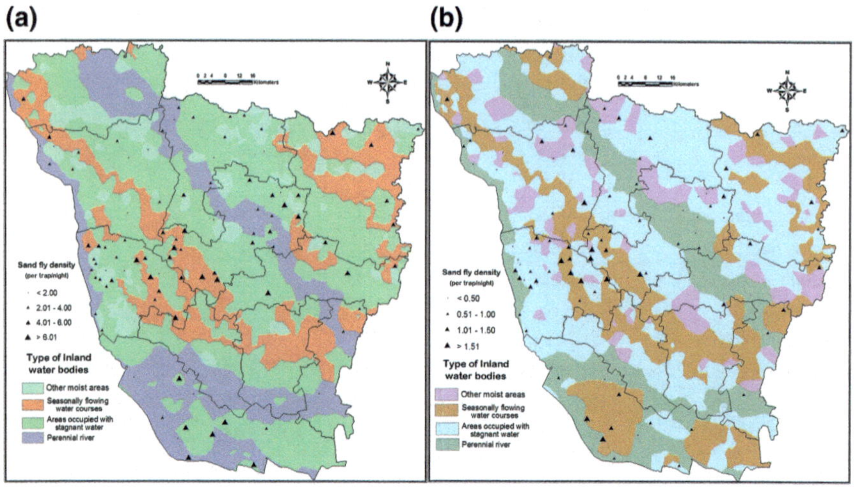

Fig. 5.11 Location-allocation analysis of seasonal inland surface water body in relation to vector abundance of Vaishali district, **a** lean season, **b** peak season

Table 5.4 Weights for the seven themes for potential sand fly abundance mapping in Vaishali district

Environmental variables	Weight in lean season	Weight in peak season
Inside room temperature (IRT)	4	3.5
Inside room humidity (IRH)	4.5	5
Land surface temperature (LST)	3	4
Re-normalized difference vegetation index (RDVI)	3.5	3.5
Wetness index (WI)	3	4
Land use/land cover (LULC)	3.5	3.5
Location-allocation analysis of seasonal water body (LASWB)	2	2.5

5.5.3.8 Weight Assignment and Static-Prescriptive Model Analysis

Suitable weights were assigned to the seven themes and their individual features after understanding their spatial and statistical relationship with the environmental features in relation to the abundance of *P. argentipes* in the study area (Table 5.4). The normalized weights of the individual themes and their different features were obtained through the Saaty analytical hierarchy process (AHP). The weights assigned to different themes in the peak and lean season are presented in Table 5.3. Tables 5.5 and 5.6 showed the pairwise comparison results of the adaptation

Table 5.5 Pair-wise comparison matrix of seven environmental layers in lean season of Vaishali district

Environmental variables	IRT	IRH	LST	RDVI	WI	LULC	LASWB	Nth root of the product	Normalized weight	Priority weight
IRT	1.00	0.89	1.33	1.14	1.33	1.14	2.00	1.27	0.17	1.18
IRH	1.13	1.00	1.50	1.29	1.50	1.29	2.25	1.45	0.20	1.33
LST	0.75	0.67	1.00	0.86	1.00	0.86	1.50	0.91	0.12	0.89
RDVI	0.88	0.78	1.17	1.00	1.17	1.00	1.75	1.08	0.15	1.03
WI	0.75	0.67	1.00	0.86	1.00	0.86	1.50	0.91	0.12	0.89
LULC	0.88	0.78	1.17	1.00	1.17	1.00	1.75	1.08	0.15	1.03
LASWB	0.50	0.44	0.67	0.57	0.67	0.57	1.00	0.56	0.08	0.59

Consistency ratio (CR) = 0.007

Table 5.6 Pair-wise comparison matrix of seven environmental layers in peak season of Vaishali district

Environmental variables	IRT	IRH	LST	RDVI	WI	LULC	LASWB	Nth root of the product	Normalized weight	Priority weight
IRT	1.00	0.70	0.88	1.00	0.88	1.00	1.40	0.95	0.13	0.94
IRH	1.43	1.00	1.25	1.43	1.25	1.43	2.00	1.45	0.20	1.34
LST	1.14	0.80	1.00	1.14	1.00	1.14	1.60	1.11	0.16	1.07
RDVI	1.00	0.70	0.88	1.00	0.88	1.00	1.40	0.95	0.13	0.94
WI	1.14	0.80	1.00	1.14	1.00	1.14	1.60	1.11	0.16	1.07
LULC	1.00	0.70	0.88	1.00	0.88	1.00	1.40	0.95	0.13	0.94
LASWB	0.71	0.50	0.63	0.71	0.63	0.71	1.00	0.64	0.09	0.67

Consistency ratio (CR) = 0.0009

options for the environmental layers with respect to the *P. argentipes* abundance. Factor weights were determined by pairwise comparison of seven environmental themes and the consistency ratio was 0.007 in the lean season; while in the peak season, the consistency ratio was 0.0009.

5.5.3.9 Final Suitability Map of *P. argentipes* Abundance Based on Static-Prescriptive Model

- *Lean season*

Figure 5.12 shows the probable sand fly abundance zone of Vaishali district in lean season based on the static-prescriptive model using a Saaty's AHP method. The resultant value of probable *P. argentipes* abundance calculated through AHP method was divided into five distinct zones, like (i) very less abundance (less than 45.96), (ii) less abundance (45.97–55.57), (iii) moderately abundance (55.58–64.07), (iv) high abundance (64.08–73.13) and (v) very high abundance (more than 73.14). However, the very high density zones were portrayed mainly in the western part of the district, and some small zones were also delineated northeast, southwest and central-east of the district. On the contrary, the less *P. argentipes* abundance zones were also depicted in the southeast, central and north-west corner of the

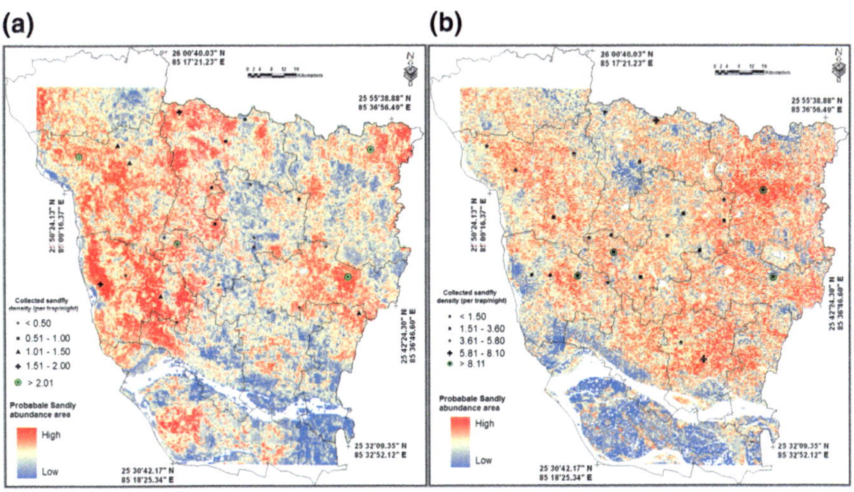

Fig. 5.12 Probable sand fly abundance map derived through Saaty's Analytical Hierarchy Process (AHP) of Vaishali district, **a** lean season, **b** peak season

district. In the southern part, the major portion of the district exhibits less to very less abundance of *P. argentipes* probability.

- *Peak season*

Probable *P. argentipes* abundance map of Vaishali district in the peak season is shown in Fig. 5.12. The map of the study area has been divided into five zones based on the probability index value derived through Saaty's AHP method, as follows: (i) very less abundance (less than 58.54), (ii) less abundance (58.54–68.53), (iii) moderately abundance (68.53–76.93), (iv) high abundance (76.94–85.08) and (v) very high abundance (more than 85.09). The result of our analysis showed that the probability of *P. argentipes* abundance distributed across the entire district in the peak season, except the southern part. Very small parts of the southern pockets have some high probability value for *P. argentipes* abundance. However, the maximum probability value (shown 'red' colour in the map) is recorded through the eastern part of the district. There are some parts of the district where probability value is very low or nil; it may be due to the allocation of permanent surface water body that was marked as 'restricted' for *P. argentipes* habitat during the analysis.

5.5.3.10 Model Comparison and Validation

Table 5.7 summarizes and evaluates the probable *P. argentipes* abundance models, presents the abundance approximation and substantiation results. In the lean season, out of 25 sample sites, 15 sample sites were correctly matched with our derived result. Conversely, in the peak season, only 13 sample sites were correctly matched with the final outcome. The overall accuracy of the model in lean season showed 60.00% accuracy; while, in the peak season, the model accuracy was 52.00%. It may be due to the small number sample sites were used to estimate the results. It appears, nevertheless, that *P. argentipes* abundance might be tricky to forecast and that prospective modeling efforts should consider additional factors such as socioeconomic and demographic indicators if address-specific environmental data are available.

5.5.4 Discussion

Spatial estimation of *P. argentipes* abundance in Vaishali district was carried out by analyzing environmental data to develop habitat suitability models. However, the analysis has been carried out in two different seasons (e.g., peak and lean), and still little is known about the vector resting places during lean season and the process of re-emergence in peak season. However, in the present work, only seven environmental variables, like indoor climate (e.g., temperature and humidity), land surface temperature, re-normalized difference vegetation index, wetness index, land use/land

Table 5.7 Error matrix table for model comparison and validation

Class density	Reference totals	Class density information	Number correct	Producer accuracy	User accuracy
Lean season					
<45.96	9	4	5	55.56	100.00
45.97–55.57	7	6	3	42.86	50.00
55.58–64.07	4	5	3	75.00	60.00
64.08–73.13	2	7	2	100.00	28.57
>73.14	3	3	2	66.67	66.67
Total	25	25	15		
Overall model accuracy 60.00%					
Peak season					
<58.54	5	6	2	40.00	33.33
58.54–68.53	6	6	3	50.00	50.00
68.53–76.93	9	7	5	55.56	71.43
76.94–85.08	2	3	1	50.00	33.33
>85.09	3	3	2	66.67	66.67
Total	25	25	13		
Overall model accuracy 52.00%					

cover and seasonal water body were considered to map the potential *P. argentipes* abundance zone.

Like many other vectors, *P. argentipes* the vector of Indian Kala-azar is greatly influenced by the local environmental factors in India (Ready 2008; Bhunia et al. 2012a, b, c). Some authors advocate that remotely sensed environmental data and ground based observation of meteorological data allowed the prediction of adult *P. argentipes* densities (Kesari et al. 2011; Kassem et al. 2012; Gálvez et al. 2010). Presently, we focused to estimate the association between the geo-environmental factors and *P. argentipes* abundance, and all of these factors have been measured together as a composite index to delineate the suitable area for vector abundance using Saaty's AHP process. The static-prospective methods of variable weighting often used in GIS analyses can initiate individual bias in the modeling process. The principle of AHP is to methodically break down a difficulty into its smaller and smaller ingredient parts and then conduct decision makers through a series of pairwise comparison conclusions to articulate the significance of the elements in the hierarchy (Saaty 2005; Bana e Costa 2005). The method can also be worn for group decision making—that is imperative for this study—since diverse point of view of health planners can be included into the hierarchical composition of this model.

The vector is effectively a 'wet season' species and with the arrival of rains there is usually a spiky upswing in the density of the species. Previously, it was evidence that the sand fly started building up in the post monsoon season, when mean temperature varied from 27.5 to 31.0 °C and relative humidity 73–93% (Bhunia et al. 2010a, b; Sharma and Singh 2008). Delimitation of spatial and temporal changeability in LST values, for example, may be used as a correlative index of vector abundance (Malone et al. 1994; Rogers et al. 1996). *P. argentipes* abundance was low and negligible when the temperature increases and/or decreases. Our results also documented that mean and maximum LST has a noteworthy effect for *P. argentipes* abundance in the lean season. Thus, it is recommended that the utility of LST in *P. argentipes* abundance mapping may be significantly enhanced in studies of epidemiological research, especially for Kala-azar transmission. Mean RDVI values are extremely valuable and effective in analyzing the conditions of *P. argentipes* abundance. Correspondingly, there are frequently robust associations of disease and vector abundance with the amount and density, rather than the species composition of vegetation cover (Rejmankova et al. 1991; Thomson et al. 2003).

Analysis of LULC features revealed that adult *P. argentipes* density was significantly associated with land cover variables (e.g., settlement, surface water body, moist fallow, vegetation, sand and river). In our analysis, it is observed that the class density of agricultural land is very less and had very less influence on the abundance of *P. argentipes*. It may be due to the fact that the agricultural areas around Vaishali district could be related to mechanical or chemical devastation of sand flies breeding sites (Dedet and Pratlong 2008; Faucher et al. 2012). The importance of surface water bodies liesin the fact that these contribute to maintain soil moisture conditions at the sub-soil surface, which in turn ensembles breeding proliferation of immature stages of sand fly as well as adult resting niche. In an earlier report, conducted by Sudhakar et al. (2006) and Kesari et al. (2011) demonstrated a significant correlation of vector density with variables like temperature; humidity and landscape environment, which stalwartly, supported our study. Similarly, human settlement has also been conscientious with the existence of the sand fly, because humans co-operate a function infeeding of their preferred host and also act as a reservoir (Rahman et al. 2010). In this area, small patches of banana/bamboo/mango/litchi plantation were also found surrounding the houses, which are dominant cash crop for local people, aids to the abundance of *P. argentipes*. Because, these phlebotomine sand flies, like other biting flies, require sugar for survival and several different sources of sugar meals of insect origin (honeydew), and of plant origin, have been identified (Schlein and Warburg 1986; Moore et al. 1987; Müller and Schlein 2006).

WI (Tassled cap transformation) is a measure of the moisture in soils, vegetation, and other surface cover (Crist et al. 1986; Crist and Cicone 1984), is valuable for characterizing such biophysical spatial patterns connected with *P. argentipes* habitat suitability. Because, soil moisture is an important factor for survival of sand fly (Singh et al. 2008a, b). Bhunia et al. (2011) scrutinized that the presence of streams and other water bodies plays an important role in the distribution of vector as well as affect the Kala-azar incidence.

However, our study suggests that these environmental factors are important for the successful determinants of Kala-azar vector (*P. argentipes*) in Indian sub-continent. The predictive value of this remote sensing map based on indoor climate, LST, RDVI, WI, LULC and location-allocation of inland surface water bodies appears to be better for the forecast of the *P. argentipes* habitat areas. The probable seasonal sand flies abundance map was validated with the current seasonal sand fly abundance data, and clearly shows areas environmentally prone to sustaining sand fly profusion. Furthermore, the methodology developed in this research is straightforward and can be effortlessly tailored for this and other vector-borne diseases in different ecological regions. This static risk map can voluntarily be rehabilitated to a dynamic web-based information delivery system where selecting an area of interest supplies a close-up view showing risk patterns for labeled spatial units.

5.6 Case Study 2: Spatial Correlation of Climatic and Environmental Factors of Visceral Leishmaniasis of Bihar, India: A Geoinformatics Approach

5.6.1 Introduction

Visceral leishmaniasis (VL) or Kala-azar, caused by protozoan parasites *Leishmania donovani* and transmitted by the vector, *Phlebotomous argentipes*, has existed a large burden of disease in Indian sub-continent targeting the poor (WHO 2007). It is a second most prevalent parasitic disease ranking after malaria. In Bihar, it was first recorded as "Kala—dukh" at Purnea district in 1882 (Bora 1999). Recently, the disease is endemic only in Bihar, West Bengal, Uttar Pradesh and Jharkhand. Large-scale control programme, mainly based on household spraying of residual insecticides, has achieved cost effective local eradication of domestic vectors and interruption of *P. argentipes* in northern part of the continent (Kumar et al. 2015; Thakur 2007). Bihar state contributes 50% of the VL caseload in the subcontinent and 90% in India (Joshi et al. 2006). However, the figure from the northern districts of Bihar is astonishing, with 37,738 and 29,078 cases in 2007 and 2009 respectively (Report from State Health Society of Bihar, India, 2007).

VL is associated with multiple factors, including environmental changes that affect the population biology, development, and behavior of vectors, as well as dimensions that determine the population biology, and even of behavior of human. Meteorological factors (i.e. temperature, rainfall) and environmental factors (i.e. soil temperature and moisture) have been associated to *P. argentipes* monthly abundance in Bihar (Picado et al. 2010a, b, c) and West Bengal (Ghosh et al. 1999) respectively. The spatial distribution of VL is markedly heterogeneous, which may lead to a substantial increase in transmission levels (Werneck 2008). In this situation, focusing intervention on high risk areas might be an efficient strategy to reduce transmission rates.

During the last decades, remote sensing has been applied to a broad range of fields. In recent time satellite, data are used to obtain a variety of types of geographical and landscape information (Beck et al. 2000; Hay et al. 1996). Across continental extents and broad areas, environmental factors like humidity, temperature, rainfall and land cover features highly be influenced the distribution and developments of *Phlebotomous argentipes* development that can limit VL occurrence (Bhunia et al. 2010a, b; Bucheton et al. 2002; Ranjan et al. 2005).

The main objective of the present study is to characterize and measure the association between different climatic and environmental factors to risk assessment for VL at endemic and non-endemic district of Bihar. The VL cases in 2007–2009 across the Vaishali district in North Bihar and Gaya district in South Bihar were associated with environmental factors describing aspects of surface reflectance and topography.

5.6.2 Objectives

In order to achieve the main objective, the following specific objectives are identified to: (i) study the association between different climatic and environmental factors in relation to distribution of VL; (ii) identify/map the risk prone areas for quick remedial measures; and (iii) assess the potential of remote sensing and GIS for rapid assessment of land use/land cover and their relation with the VL leading to the mapping of risk prone areas.

5.6.3 Materials and Methods

Study area

The study was conducted at district level of Bihar state in India i.e. Vaishali district (25° 29′ N to 26° 00′ N and 85° 04′ E to 85° 38′ E), as Kala-azar is highly endemic in this area and Gaya district (24.50° N–25.10° N and 84.40° E–85.50° E), selected as control area for monitoring sand fly density (Fig. 5.13), as there are no cases of VL from the last decade.

The endemic district enjoys a bracing and healthy climate with maximum and minimum temperature of 36.9 and 5 °C in the summer (April–May) and winter (December–January) months respectively, whereas, annual average rainfall of the district approximately 1,170 mm, which is recorded during the monsoon period (June–July). Non-endemic district suffers from extremes of climate, attributable to the barren rocky hills and vast sandy bed of the river Phalgu to its immediate east. The maximum and minimum temperature of the district is 46 and 4 °C in the month

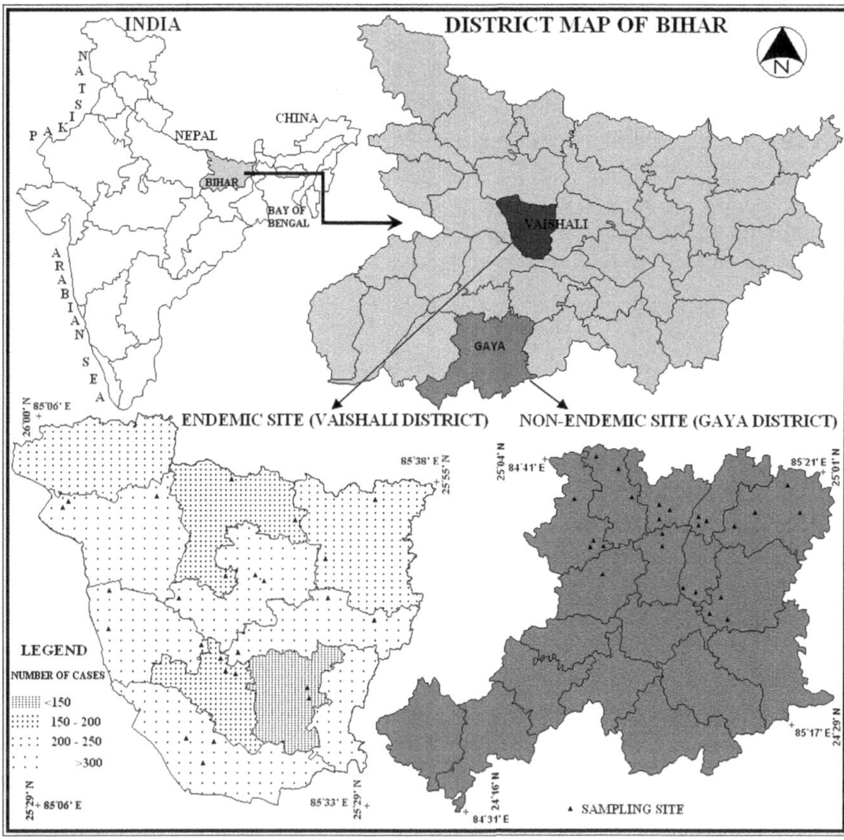

Fig. 5.13 Map showing the study area (endemic and non-endemic)

of May and December respectively; and an annual average rainfall in the district is 1,150 mm. As such, both the endemic and non-endemic districts representing different type of ecosystem.

Entomological and epidemiological data

Adult sandflies were monitored randomly in households and/or cattle sheds from the each district between January and December 2007 using Communicable disease Centre (CDC) light traps placed at 50–70 cm above the ground, and at 1 feet far from the wall. Traps were run (between 18.00 and 06.00 h) once a month, early in (in the first week of) each month. The counts of *P. argentipes* in the traps were used to calculate mean monthly numbers of this sandflies/trap-night, as a measure of the density of the local vector population. However, the density of sand fly was collected only at the northern part of the district.

Government agencies provided data on VL cases on 2007 and diagnosis was based on parasite presence in spleen/bone marrow aspirations, identified with rk39 kits as per the protocol provided by National Vector borne Disease Control Programme (NVBDCP), government of India.

Climatic data

Data on mean monthly temperatures, including maximums and minimums, humidity, sea level pressure (SLP), precipitation as well as mean wind velocity for 2007 was collected from Indian Meteorological department's (IMD station) weather stations (Patna and Gaya) on the endemic and non-endemic site of Bihar, respectively.

Satellite data processing and interpretation

The potential risk areas for VL cases occurrences were estimated based on indirect, landscape scale measures that are correlates of environmental factors associated with the disease-vector interactions that make up the etiology of this disease. In particular, the focus was on correlates of three key environmental dimensions: land use/land cover features, vegetation and topography. The Landsat 4–5 Thematic Mapper (TM) and Advanced Spaceborne Thermal Emission and Reflection Radiometer (ASTER) satellite data were used to extract these environmental features. The details of satellite data was provided in Table 5.8.

The satellite data was geo-referenced with respect to the known reference points, obtained through Global positioning system (GPS) surveys during the field visit. The entire scenes were registered to a UTM co-ordinate system (Zone 45 N and datumWGS-84) based on second order polynomial algorithm and nearest neighbor resampling method.

Land use/land cover maps were generated to identify the different classes of land from Landsat5 TM imagery. A supervised classification technique with Maximum Likelihood (MXL) algorithm was used to assign the pixel into eleven land cover classes for endemic site and ten classes for non-endemic site, based on their spectral reflectance characteristics. After selectively combining classes, classified images were sieved, clumped and filtered before producing final output. Sieving remove isolated pixels using blob grouping, while clumping helps maintain spatial coherency by removing unclassified black pixels (speckle holes) in classified images.

Normalized Difference Vegetation Index (NDVI) is a numerical quantity derived from reflectance measured in the red and near-infrared spectral bands that provides

Table 5.8 Characteristics of satellite images used for the study

Satellite data	Retrieval type	Spatial/spectral resolution	Date of acquisition (path/row)	Source of data
Landsat5 TM	GeoTIFF	28.5 m (B, G, R, NIR) 30 m (MIR) 120 m (TIR)	2006-10-30 (141/042) 2006-10-30 (141/043)	www.landcover.org
ASTER	GeoTIFF	30 m	2009	http://www.gdem.aster.ersdac.or.jp

information about photosynthetic activity (Tucker 1979). The value of NDVI can be calculated as follows:

$$\mathrm{NDVI} = \mathrm{NIR} - \mathrm{RED}/\mathrm{NIR} + \mathrm{RED}$$

The ASTER Global Digital Elevation Model (GDEM) is produced with 30 m postings, and is formatted in 1 × 1 degree tiles as GeoTIFF files (http://www.gdem.aster.ersdac.or.jp). Finally, the elevation is validated based on benchmark of Survey of India (SOI) topographical sheet (scale—1:50,000).

Statistical analysis

Data was analyzed using statistical software Stata ver10. Month variable was transformed into season as ordinal variable considering the lowest vector density in winter and highest during September–October month. Student's t-test was applied to compare the difference of mean sand fly density in endemic and non-endemic sites. In order to estimate the effect of climatic variables on sand fly density, mainly *P. argentipes* density, multivariate linear regression analysis was carried out using backward step method thus putting all the observed independent variables into the model at the same time, and removing the most insignificant variable one by one from the model until the final model is achieved.

5.6.4 Results

Sand-fly density

Monthly distribution of sand fly density in endemic and non-endemic sites is presented (Fig. 5.14). Using sand fly density, the non-endemic region had a significantly higher *P. argentipes* than the endemic regions of Bihar. The mean sand fly density of endemic site was 6.63 (95% CI, 4.28–8.99) as compared to 10.81

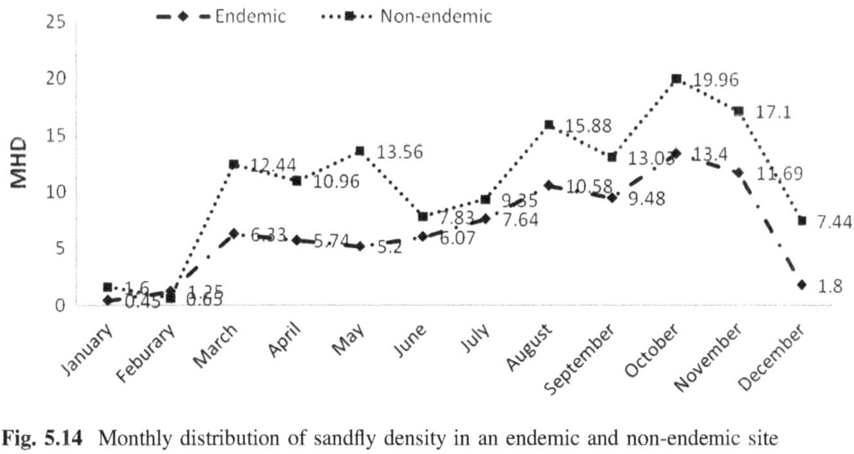

Fig. 5.14 Monthly distribution of sandfly density in an endemic and non-endemic site

(95% CI, 7.50–14.13) in the non-endemic site, showing higher sand fly density in non-endemic site as compared to endemic site but not statistically significant (p = 0.06). During this study period, aggregate population of sandflies was found to be lowest in December to February month. The sand fly population rose during March, with highest peak in October. The peaks of sand fly density were observed between August and September in both endemic and non-endemic sites.

Climatic data analysis

The recorded average monthly temperature, humidity, sea level pressure (SLP), precipitation and mean wind velocity (MWV) of 2007 for both the endemic and non-endemic sites were considered for risk analysis (Table 5.9). Considering small number of time points on which the observations were recorded (only 12 months observation), we kept all the climatic variables in multivariate regression analysis in initial stage.

Multivariate linear regression analysis using backward-step method using population of the study sites as weighted was carried out to determine predictor variables affecting sand fly density (Table 5.10). The final model used for predicting sand fly density is given by the following equation:

Table 5.9 Independent association of potential climatic variables with sandfly density in both the endemic and non-endemic sites

Climatic variable	Endemic area			Non-endemic area		
	Mean 95% CI	Correlation coeff. (r)	p-value for (r)	Mean 95% CI	Correlation coeff. (r)	p-value for (r)
Sandfly density (MHD)	6.63 (4.28–8.99)			10.81 (7.50–14.13)		
Mean temperature	26.18 (24.49–27.86)	0.49	<0.15	26.19 (24.52–27.87)	0.43	<0.15
Maximum temperature	31.98 (30.63–33.34)	0.39	<0.15	32.52 (31.16–33.87)	0.296	>0.15
Minimum temperature	20.31 (18.23–22.39)	0.59	<0.15	19.88 (17.83–21.93)	0.48	<0.15
Sea level pressure (SLP)	1007.55 (1005.56–1009.54)	−0.32	<0.15	1007.43 (1005.38–1009.47)	−0.22	>0.15
Mean humidity (MH)	67.13 (63.62–70.65)	0.35	<0.15	63.71 (58.48–68.93)	0.32	<0.15
Precipitation (PP)	78.63 (42.22–115.05)	0.34	<0.15	88.94 (46.65–131.24)	−0.03	>0.15
Mean wind velocity (MWS)	14.23 (5.24–23.21)	−0.08	>0.15	6.03 (5.29–6.78)	−0.1	>0.15

$$Y = -505.28 + (0.95 \times \text{mean temperature}) + (0.48 \times SLP) + (\text{Season} \times 0.83) + (\text{area} \times -4.22)$$

where, Y (MHD) is the estimated sand fly density.

The final model was highly significant (F = 10.99, *p*-value = 0.0001). It means that these four variables when considered together are significant predictor of sand fly density, and also the adjusted $R^2 = 0.70$, indicating that nearly 70% of the variance of sand fly density could be explained by these three predictor variables.

Table 5.10 Significant predictor variables of sand fly density

Predictor variables	Coefficients (βs) (95% CI)	SE (βs)	T-statistic	*p*-value
Intercept	−505.28 (−1019.51,8.94)	245.68	2.06	0.05
Mean temperature	0.95 (0.35, 1.54)	0.28	3.34	0.003
Sea level pressure (SLP)	0.48 (−0.01, 0.98)	0.24	2.07	0.05
Season	0.83 (0.43, 1.23)	0.19	4.33	0.001
Area	−4.22 (−7.01, −1.43)	1.31	3.17	0.005

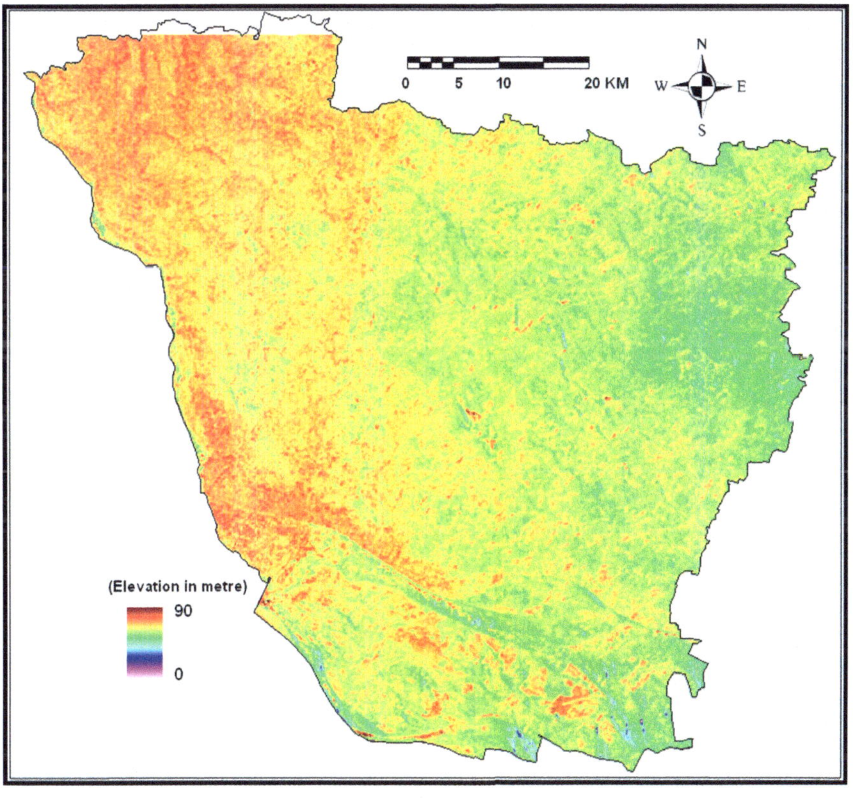

Fig. 5.15 Elevation map of endemic site (Vaishali district)

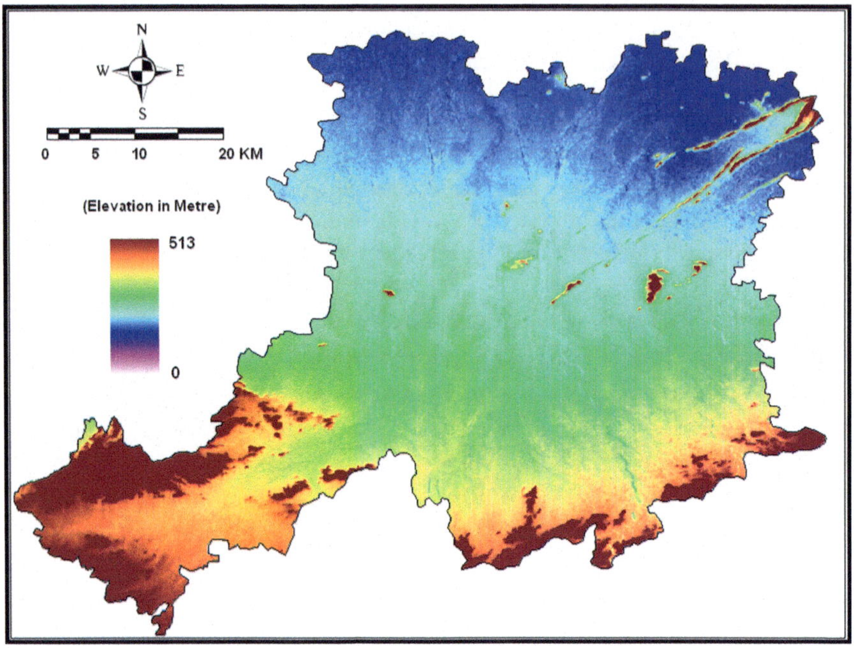

Fig. 5.16 Elevation map of non-endemic site (Gaya district)

Altitudinal factor

ASTER digital elevation model (DEM) data was used to draw the altitudinal maps of the both the study sites. Altitudinal range in an endemic site varied from 0-90 meters above mean sea level (MSL) (Fig. 5.15), whereas, in non-endemic site, it ranged from 16.03 to 513 m above MSL (Fig. 5.16). The endemic district is characterized by plain topography, covered most alluvial soil. Due to less altitudinal range, the topography of endemic region is divided into four categories with 20 meters intervals i.e. Below 40 m, 41–60 m, 61–80 m, and more than 80 m. The topographic range of an endemic site is correlated with the distribution of cases presence in an each altitudinal range. The number of cases was found to be very high at below 60 m mean sea levels (Table 5.11). At altitude of more than 60 m, very less number of cases was found, to become eventually nil.

Status of Vegetation cover

Most of the part, the region covered with edible shrubs, weeds, grass and bushes etc. As peridomestic vegetation, individual plant species such as seem, pumpkin, chillies, carrot, beat, banana, bhang, palmyra tree, date palm etc. were found. The NDVI values were computed from Landsat5 TM data for Vaishali and Gaya district. Minimum, maximum, standard deviation and average values of NDVI were

Table 5.11 Distribution of cases in endemic site (Vaishali district) at different altitude

Elevation (m)	Cases	Percent
<40	635	46.90
40–60	718	53.03
60–80	1	0.074
>80	0	0.00

Table 5.12 NDVI in the study area (computed from Landsat ETM + data)

Study site	$NDVI_{min}$	$NDVI_{max}$	$NDVI_{Mean}$	Std. deviation
Vaishali (Endemic)	−0.43	0.49	0.04	0.27
Gaya (Non-endemic)	−0.30	0.60	0.15	0.26

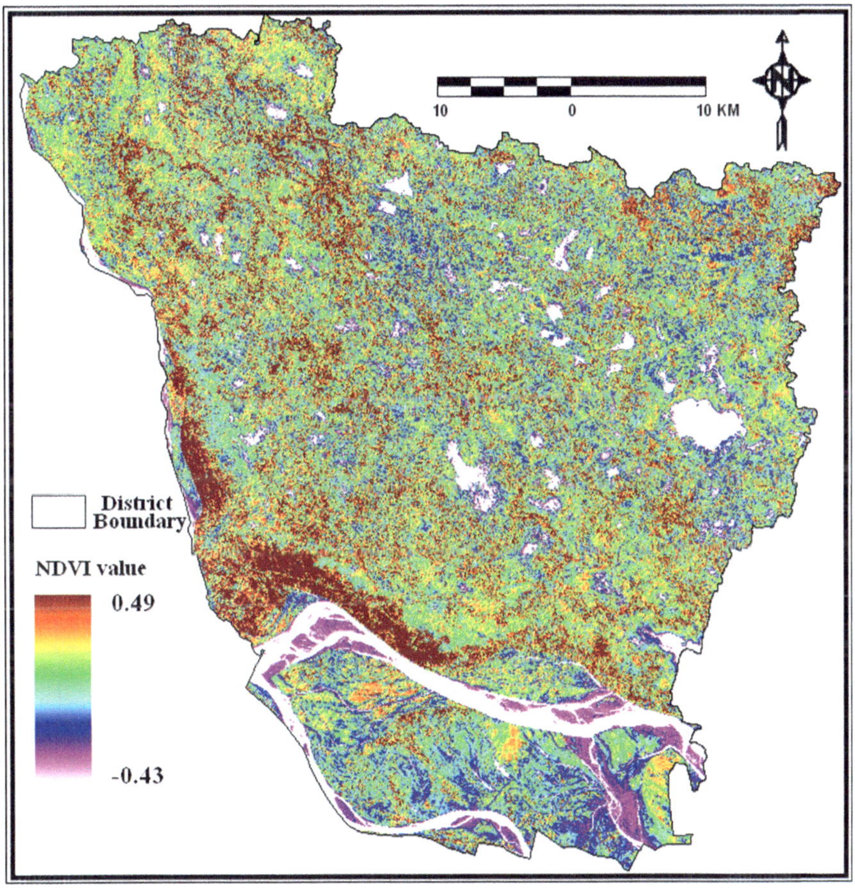

Fig. 5.17 NDVI map of Vaishali district (endemic site)

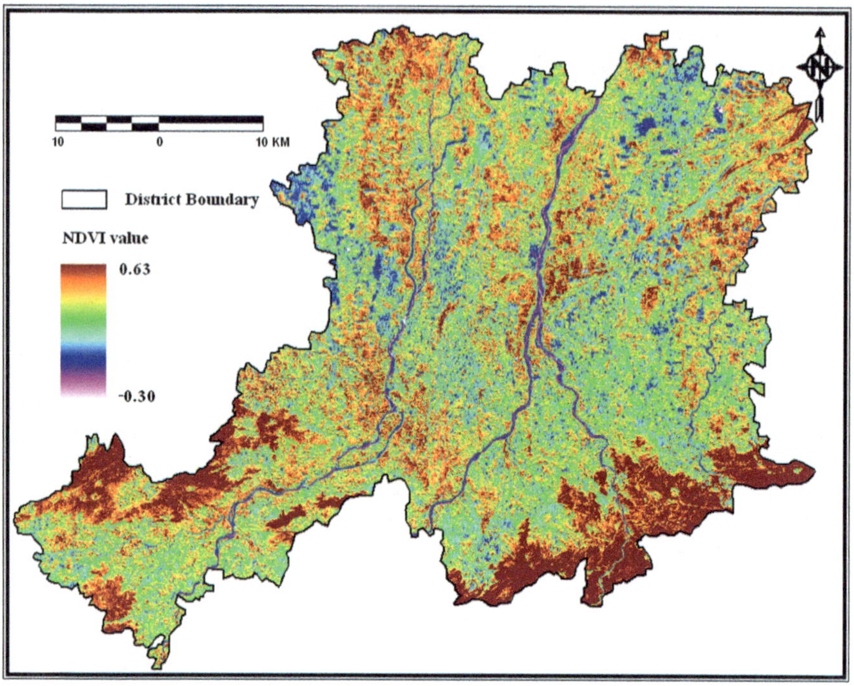

Fig. 5.18 NDVI map of Gaya district (non-endemic site)

computed (Table 5.12) and used to investigate the association between Kala-azar positives. Based on NDVI values, both these areas were classified into the vegetation density zone map (Figs. 5.17 and 5.18) which showed a variety of high density vegetation (more than 0.55), areas of medium density (0.40–0.55), low density zones (0.15–0.30) and very low density zone (less than 0.15). When we overlaid all the reported cases on the NDVI map, the results showed that most of the cases occurred in non-vegetative areas or low density vegetation zones (Fig. 5.19).

Land use/land covers characteristics

The digital maps of the study sites are presented in Figs. 5.20 and 5.21 for endemic and non-endemic site, respectively. Following land use classes were considered in image classification: river, sand, settlement, surface waterbody, marshy land, moist fallow, dry fallow, crop land, agricultural fallow, plantation/sparse vegetation, weeds/grass and dense forest in an endemic and non-endemic area (Table 5.13).

In endemic and non-endemic sites crop and agricultural land covered by 71.61% and 54.58% area, respectively. The results also illustrated that remaining 28.39% area of endemic site covered with plantation (2.53%), moist fallow (9.74%), river (3.04%), settlement (0.27%), dry fallow (0.42%), surface waterbody (1.51%), weeds/grass (6.18%) and marshy land (3.91%). In non-endemic site remaining 45.42% enclosed by dense forest (22.60%), sparse vegetation (15.85%), dry fallow

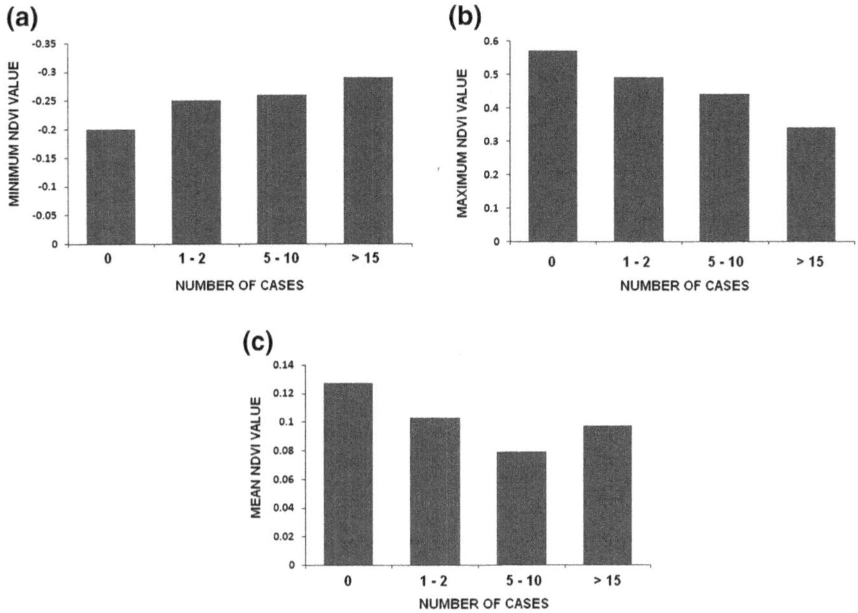

Fig. 5.19 Relation between cases distribution and **a** minimum NDVI, **b** maximum NDVI and **c** mean NDVI

(1.81%), moist fallow (1.70%), river (1.08%), settlement (1.08%) and sand (1.20%). The result also demonstrated that in an endemic site percent of surface waterbody (0.9%), marshy land (3.91%), moist fallow (8.04%) and crop land (21.94%) is much higher compare to the non-endemic site.

5.6.5 Discussion

Keeping in view the importance of identifying the factors that promote sand fly presence and its prevalence, this study was planned using remote sensing technology; a cross-sectional survey was carried out in Kala-azar endemic and non-endemic areas of Bihar. This study provides an important source of information to inform control programmes for Kala-azar disease, specifically to identify district level risk factors of Kala-azar infestation. The most important characteristics of the survey is to use of local control services and PHCs, to collect information that could be replicated in future large scale surveys and monitoring programmes.

The current investigation shows that the distribution pattern of *P. argentipes* in both the endemic and non-endemic district more or less same throughout the year. However, the density of *P. argentipes* was high in non-endemic site but not statistically significant (*P-value* = 0.06), because of occasional DDT spraying, which

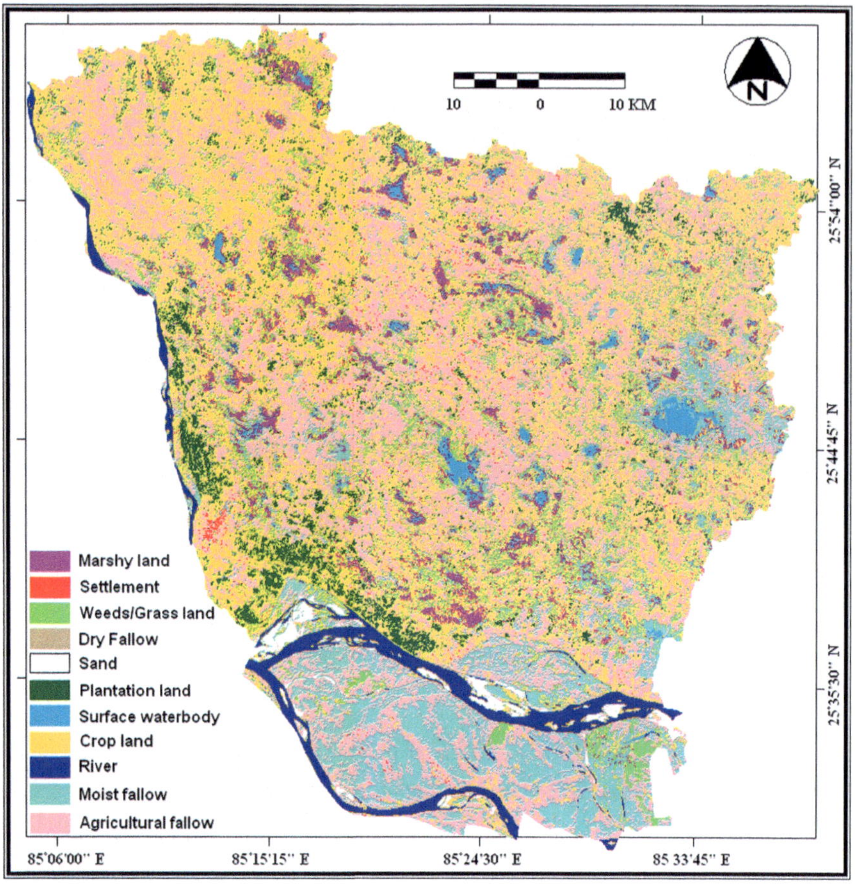

Fig. 5.20 Landuse/land cover map of endemic site (Vaishali district)

was done only on a focal basis in the event that Kala-azar occurred in an endemic form. This is consistent with the observations of Kumar et al. (2009), who found that adult *P. argentipes* in a non-endemic site of south Bihar, India (Patna, Nalanda districts) relatively high in respect to the endemic region (Vaishali and Muzaffarpur districts). Because of the high sand fly density, the susceptible populations are frequently bitten and it has been demonstrated that a salivary gland protein from the sandflies provides protection against Kala-azar infection (Valenzuela et al. 2001; Silva et al. 2005). Therefore, it appears that sand fly saliva play a crucial role in epidemiology of Kala-azar. Hence, it may be suggested that Gaya (non-endemic district) is at the zone of danger because the moment of source of infection is available from the neighboring endemic districts of Bihar and the cases may be flare off. Therefore, it is high time to take precautionary measure to prevent the epidemic.

Climate of any region is chiefly determined by such meteorological influences as relative humidity of air, temperature, wind, SLP. Analysis of the climatic variables

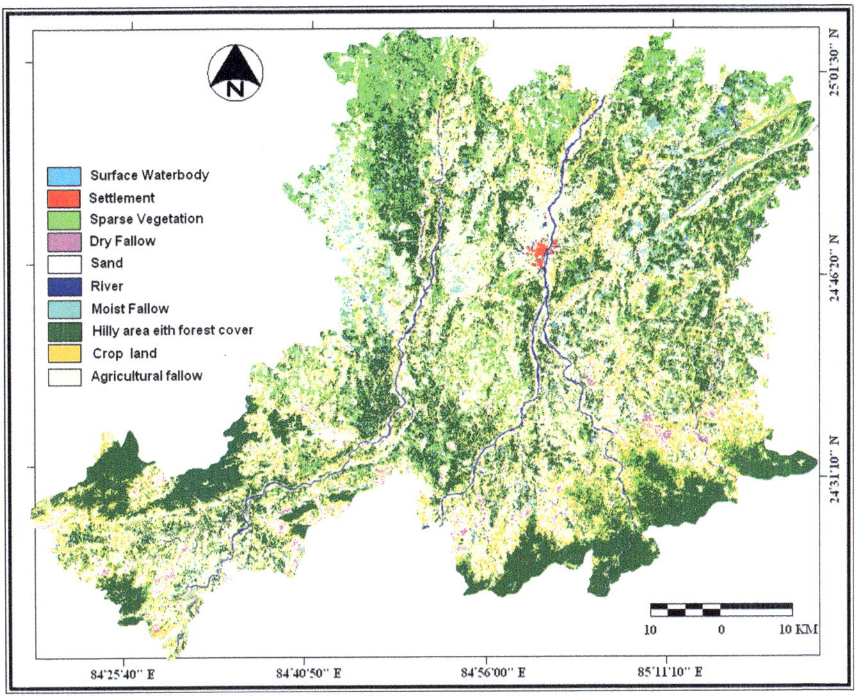

Fig. 5.21 Land use/land cover map of non-endemic site (Gaya district)

indicates statistically robust relationship between climatic parameters with the density of sand fly, *P. argentipes*. Since the behavior of the sandflies directly influenced by the temperature conditions, lower temperature should reduce their

Table 5.13 Land cover features in Vaishali and Gaya district (computed from Landsat5 TM data)

Land use/ Land cover variables	Description	Vaishali District (Endemic)		Gaya District (Non-endemic)	
		Area (Hect.)	Percent	Area (Hect.)	Percent
River	Natural water course; usually fresh water	6141.78	3.04	5387.76	1.08
Sand	Naturally occurring granular material at the river basin	1601.91	0.79	5977.8	1.20
Settlement	High residential area	549.72	0.27	2765.91	0.56
Surface waterbody	Includes 'pond', 'Chaur' and low laying wet land areas	3044.52	1.51	3058.88	0.61
Marshy land	Land covered with shallow bodies of water and aquatic vegetation	7885.89	3.91	–	–
Moist fallow	Land saturated with water	19647.5	9.74	8471.97	1.70

(continued)

Table 5.13 (continued)

Land use/ Land cover variables	Description	Vaishali District (Endemic)		Gaya District (Non-endemic)	
		Area (Hect.)	Percent	Area (Hect.)	Percent
Dry fallow	Area characterized with no green vegetation, unusable and unsaturated land	846.27	0.42	9021.41	1.81
Crop land	Characterized by seasonal crops, intensively managed for production of food	74271.6	36.81	74121.60	14.87
Agricultural fallow	Areas dominated by shurb land and do not exhibit visible vegetation as a result of being tilted in a management practice	70213.3	34.80	197922.0	39.71
Plantation/ Sparse Vegetation	Areas dominated by non-natural woody vegetation	5101.02	2.53	78981.7	15.85
Forest	Areas characterized by tree cover; canopy accounts for 25–75%	–	–	112648.29	22.60
Weed/Grass	Areas with grass covered/pasture land	12471.8	6.18	–	–
Total area		201775.3		498357.3	

activity and extreme values would completely inhabit their movement (Killick-Kendrick 1983). All the metabolic process of sand fly is influenced by temperature (Benkova and Volf 2007); and affects the fecundity of sand fly (Ferro et al. 1998; Kaburi et al. 2011; Theodor 1936). In the present study, mean temperature build a strong relationship with the sand fly density. However, there was no observable variation in the monthly distribution of maximum and minimum temperature of endemic and non-endemic site. In case of relative humidity, no statistical significant difference was found in an endemic and non-endemic site. In previous study, it was found that relative humidity plays a significant role in their life cycle (Singh et al. 2008a, b), but in this present study no statistical significant relationship was found between sand fly density and mean humidity of endemic and non-endemic site. Sea level pressure (SLP), is the pressure at sea level, influence the local climate of a region (Dunlop 2003; Williams 1997). Lowering the air temperature increases its relative humidity for the same partial pressure of water vapor. In general, relative humidity of a parcel of air will increase as the air rise, because the temperature effects outweigh the sea level pressure effect. In this study, SLP shows a strong positive relationship with the sand fly density. In both the site, there is no significant liaison between the vector density and mean wind velocity. Overall, a multivariate linear regression model was derived from the climatic variables, shown that mean temperature, SLP, season and area would be useful to produce a predictive model for the distribution of *P. argentipes* and can be used to predict the vector density in those areas not covered by the initial data. Hence, the climatic variables, which were found significantly associated with sand fly density, might have independent influence on vector population for their breeding and propagation (Anon 1993; Napier 1926).

The general abundance of Kala-azar cases was positively correlated with the minimum, mean and maximum NDVI value in an endemic site. The maximum NDVI value was significantly higher in non-endemic site compared to the endemic site. This result indicated that low vegetation cover area (minimum NDVI) is highly sensitive for transmission of Kala-azar, because sandflies requires sugar meal, taken from plant material (Said 1986; Schlein and Jacobson 1999); therefore they would be expected to be less abundant in areas with little peridomestic or soft stem vegetation. This soft stem peridomestic vegetation assists pierce and imbibe plant sap by male sand fly population and may supply as a source of sucrose for naturally infected sandflies for development and multiplication of *Leishmania Spp.* within its gut (Sudhakar et al. 2006). Several studies have shown that the most of the Kala-azar cases occurred in non-vegetative areas or low-density vegetation zones in Indian (Bhunia et al. 2010a, b; Cross et al. 1996) and Sudan (Elnaiem et al. 1997, 2003).

It is also obvious from the DEM results that the altitude has a relatively strong influence on the distribution of *Phlebotomus* vector, as the Kala-azar incidence is concentrated at low altitudes. Similar work by Hermeto et al. (1994), Raina et al. (2009) and are also supported by Bhunia et al. (2010a, b) who investigated the influence of altitude on the incidence of Kala-azar in eastern Sudan and Indian sub-continent. The trend observed indicates that a high incidence should be expected at lower elevation, is probably due to its effect on temperature. Lower temperature at high altitudes will reduce the development and survival of larval stages and slow down the development of the parasite in the vector, reducing the chance of Kala-azar transmission (Benkova and Volf 2007; Zilberstein and Shapira 1994). It seems thus clear that altitude is an important environmental factor that must be taken into account when attempting to predict the potential for Kala-azar outbreaks.

The land-use/land-cover classification of the study area indicated that most (more than 50%) of the total area of both sites were covered by crops and agricultural land. The remaining—where almost all of the new cases of VL recorded in the study area—covered by river, plantation, settlement, surface water body, marshy land, dry fallow, moist fallow and weeds/grass. Chiefly, the conditions under which the vectors can thrive are limited, human visceral leishmaniasis appears to be a disease that is very sensitive to environmental characteristics (Raina et al. 2009; Elnaiem et al. 1998; Victora et al. 1997). The main goal of the present study was to determine if those environmental conditions observable via TM imagery were correlated with the local abundance of *P. argentipes* and thus with the level of transmission of *L. donovani* and, consequently, the incidence of VL. Analysis of land use/land cover features revealed that adult sand fly density was positive and higher for all the study sites of endemic and non-endemic focus, wherein at least

50% land should have crop and agricultural fallow land. The importance of these land cover features lies in fact that these contribute to maintain soil moisture conditions in soil/sub-soil level, which in turn suits breeding propagation of immature stages of sandflies as well as adult resting habitats. In a previous study, Sudhakar et al. (2006) demonstrated a significant correlation of vector density with variables like temperature, humidity and land use/land cover characteristics, which is strongly, supported our study.

5.6.6 Conclusion and Outlook

Spatial heterogeneity in the incidence of Kala-azar is an important aspect to be considered in planning control actions for the disease. The current study clearly brought out the usefulness of remote sensing and GIS in generating the crucial information on spatial distribution of land use/land cover classes with special emphasis on indicator land cover classes thereby helping in prioritizing the area to identify risk prone areas of Kala-azar through GIS application tools. Climatic variables (mean temperature, SLP) were found significantly associated with sand fly density. Analysis of land use/land cover features revealed that adult sand fly density was positive and higher for all the study sites of endemic and non-endemic focus, wherein at least 50% land should have crop and agricultural fallow land. Favorable condition for the sand fly breeding in an area is arising due to specific environmental composition. Some clear-cut links- between endemic VL and climatic parameters (e.g. mean temperature, SLP, season), certain ranges of altitude (<60 m above MSL), minimum NDVI and land cover features (e.g. surface waterbody, swampy land, settlement, agricultural fallow land etc.) should be useful to assess for macro stratification for vector abundance/ non-abundance vis-à-vis Kala-azar occurrence/non-occurrence area, at least in India.

This study has provided a foundation, which serves as a guide for researchers in the management of different studies in a variety of aspects of Kala-azar in this region. Further analysis will investigate others environmental determinants of the geographic distribution of vectors, as well as the interrelationships between measures of vector infestation and infection risk in humans, in order to optimize allocation of surveillance and control effort. Based on the paper results, there has been a surge of interest for future research to understand the ecological and environmental parameters surrounding the sand fly of VL. Understanding these boundaries can help formulate an Early Warning System that can help identify and forecast conditions that increase the likelihood of transmission.

5.7 Conclusion

At the present time, almost all sectors make practice of GIS solicitations particularly for their planning activities. Epidemiological applications are one of these tactics and this existing growth encompasses the usability of GIS in health sciences. Now a day, most public health glitches are painstaking in the scope of bearable development and GIS is a proficient tool to improve viable development plans with its comprehensive arrangement processing in a system methodology. It is palpable that the use of GIS in health geography field will efficiently aid to progress and outspread the strategic plans of wellbeing and healthcare services. Investigating with medical doctors by sharing the involvements associated fields will also enhance a new hallucination to GIS experts and it will be conceivable to cultivate enhanced works through these practices.

GIS technology assists the epidemiologists to comprehend the phenomenon effortlessly and inhibit in it quickly. Moreover, GIS offers a better understanding on epidemiological problems by presenting statistical data in a spatially referenced way by using maps. In this way, results of the studies can easily be opened to the public to make their conscious of these kinds of subjects.

Consequently, it will be possible for the epidemiologists to:

- determine geographic distribution of disease,
- surveillance and control the infectious diseases,
- determine the epidemics, to monitor their expansion and to decide intervention strategies,
- analyze spatial and temporal trends of diseases, and
- determine the populations at risk by creating risk maps by using GIS.

References

Adda J (2015) Economic activity and the spread of viral diseases: evidence from high frequency data. IZA DP No. 9326. ftp.iza.org/dp9326.pdf

Adler S, Theodor O (1957) Transmission of disease agents by phlebotomine sand flies. Ann Rev Entomol 2:203–225

Anon (1993) Proceedings of Workshops on Entomological and Vector Control of Kala-azar. Delhi: National Institute of Communicable Diseases (Directorate General of Health Services)

Artis DA, Carnahan WH (1982) Survey of emissivity variability in thermography of urban areas. Remote Sens Environ 12:313–329

Bana e Costa CA, De Corte JM, Vansnick JC (2005) On the mathematical foundation of MACBETH. In: Multiple criteria decision analysis: state of the art surveys. In: Figueira J, Salvatore G, Ehrgott M (eds) Springer, Berlin Heidelberg: New York, NY, USA, pp 409–442

Beck LR, Lobitz BM, Wood BL, Wood LW (2000) Remote sensing and human health: new sensors and new opportunities. Emerg Inf Dis 6:217–226

Belward AS, de Hoyos A (1987) A comparison of supervised maximum likelihood and decision tree classification for crop cover estimation from multitemporal LANDSAT MSS data. Int J Remote Sens 8(2):229–235

Benkova I, Volf P (2007) Effect of temperature on metabolism of *Phlebotomus papatasi* (Diptera: Psychodidae). J Med Entomol 44(1):150–154

Bethesda: National Institutes of Health (US) (2007) Understanding emerging and re-emerging infectious diseases. http://www.ncbi.nlm.nih.gov/books/NBK20370/

Bhunia GS, Chatterjee N, Kumar V, Mandal R, Das P, Kesari S (2012a) Remote sensing and GIS: tools for the prediction of epidemic for the intervention measure. India Geospatial Forum 2012–14th annual international conference and exhibition on geospatial information technology and applications, held on 7–9 February, 2012 at Epicentre, Apparel House, Gurgoan, India

Bhunia GS, Chatterjee N, Kumar V, Siddiqui NA, Mandal R, Das P, Kesari S (2012b) Delimitation of Kala-azar risk areas in the district of Vaishali in Bihar (India) using a geo-environmental approach. Mem Inst Oswaldo Cruz 107(5):609–620

Bhunia GS, Kesari S, Chatterjee N, Kumar V, Das P (2012c) Localization of Kala-azar in the endemic region of Bihar, India based on land use/land cover assessment at different scales. Geospat Health 6(2):177–193

Bhunia GS, Kesari S, Chatterjee N, Pal DK, Kumar V, Ranjan A, Das P (2011) Incidence of visceral leishmaniasis in the Vaishali district of Bihar, India: spatial patterns and role of inland water bodies. Geospat Health 5(2):205–15

Bhunia GS, Kesari S, Jeyaram A, Kumar V, Das P (2010a) Influence of topography on the endemicity of Kala-azar: a study based on remote sensing and geographical information system. Geospat Health 4(2):155–165

Bhunia GS, Kumar V, Kumar AJ, Das P, Kesari S (2010b) The use of remote sensing in the identification of the eco-environmental factors associated with the risk of human visceral leishmaniasis (Kala-azar) on the Gangetic plain, in north-eastern India. Ann Trop Med Parasitol 104(1):35–53

Bora D (1999) Epidemiology of visceral leishmaniasis in India. Nat Med J India 12(2):62–68

Brown JD, Swayne DE, Cooper RJ, Burns RE, Stallknecht DE (2007) Persistence of H5 and H7 avian influenza viruses in water. Avian Dis 51:285–289

Bucheton B, Kheir MM, El-Safi SH, Hammad A, Mergani A, Mary C, Abel L, Dessein A (2002) The interplay between environmental and host factors during an outbreak of visceral leishmaniasis in eastern Sudan. Microbes Infect 4:1449–1457

Butler JC et al (2001) Emerging infectious diseases among indigenous peoples. Emerg Infect Dis 7 (suppl.):554–555

CDC (2009) Swine Influenza A (H1N1) Infection in Two Children—Southern California, MMWR 58(15):400–402. Available at: https://www.cdc.gov/mmwr/preview/mmwrhtml/mm5815a5.htm

Centers for Disease Control and Prevention (1994) Addressing emerging infectious disease threats: a preventionstrategy for the United States. U.S. Department of Health and Human Services, Public HealthService, Atlanta

Chua KB et al (2000) Nipah virus: a recently emergent deadly paramyxo virus. Science 288:1432–1435

Chua K, Chua B, Wang C (2002) Anthropogenic deforestation, El Nin˜o and the emergence of Nipah virus in Malaysia. Malay J Pathol 24:15–21

Colacicco-Mayhugh, MG, Masuoka, PM, Grieco, JP (2010) Ecological niche model of Phlebotomusalexandri and P. papatasi (Diptera: Psychodidae) in the Middle East. Int J Health Geogr 9(2):1–9

Congalton RG (1991) A review of assessing the accuracy of classifications of remote sensed data. Remote Sens Environ 37:35–46

Connor SJ, Thomson MC, Flasse SP, Williams JB (1995) The use of low-cost remote sensing and GIS for identifying and monitoring the environmental factors associated with vector borne disease transmission. Available from: http://www.idrc.ca/...oks/285-6/index.html#page_75

Costantini C, Ayala D, Guelbeogo WM, Pombi M, Some CY, Bassole IH et al (2009) Living at the edge: biogeographic patterns of habitat segregation conform to speciation by niche expansion in Anopheles gambiae. BMC Ecol 9:16

Crist EP, Cicone RC (1984) Application of Tasseled cap concept to simulated thematic mapper data. Photogramm Eng Remote Sens 50:343–352

Crist EP, Laurin R, Ciccone RC (1986) Vegetation and soils information contained in transformed thematic mapper data. In: Proceedings of IGARSS'86 Symposium, Zurich, Switzerland, 8–11 September 1986; pp 1465–1470

Cross ER, Newcomb WW, Tucker CJ (1996) Use of weather data and remote sensing to predict the geographic and seasonal distribution of phlebotomus papatasi in Southwest Asia. Am J Trop Med Hyg 54:330–332

Daszak P, Cunningham AA, Hyatt AD (2001) Anthropogenic environmental change and the emergence of infectious diseases in wildlife. Acta Trop 78:103–116

De Sherbinin A, Carr D, Cassels S, Jiang L (2007) Population and environment. Annu Rev Environ Resour 32:345–373

Dedet JP, Pratlong F (2008) Leishmaniasis. In: Manson's tropical diseases 22nd ed. Cook GC, Zumla A, editors. Saunders Elsevier Edinburgh, pp 1341–1365

Dhima RC, Dinesh DS (1992) An experimental study to find out the source of fructose to sandflies. Indian J Parasitol 16:159–160

Dinesh DS, Dhima RC (1991) Plant source of fructose to sandflies, particularly *Phlebotomus argentipes* in nature. J Commun Dis 23:160–161

Dunlop S (2003) The weather identification handbook: the ultimate guide for weather watchers. The Lyons Press, ISBN 1585748579. pp 395–439

El Said S, Beier JC, El Sawaf BM, Doha S, El Kordy E (1986) Sand flies (Diptera: Psychodidae) associated with visceral leishmaniasis in El Agamy, Alexandria Governorate, Egypt II. Field Behav J Med Entomol 23(6):609–615

Eldridge BF (2005) Mosquitoes, the Culicidae. Biology of Disease Vectors. W.C. Marquardt. San Diego, Elsevier, pp 95–111

Elnaiem DA, Connor SJ, Thomson MC, Hassan MM, Hassan HK, Aboud MA, Ashford RW (1998) Environmental determinants of the distribution of phlebotomus orientalis in Sudan. Ann Trop Med Parasitol 92:877–887

Elnaiem DA, Hassan HK, Ward RD (1997) Phlebotomine sand-flies in a focus of visceral leishmaniasis in a border area of eastern Sudan. Ann Trop Med Parasitol 91:307–318

Elnaiem DA, Schorscher J, Bendall A, Obsomer V, Osman ME, Mekkawi AM, Connor SJ, Ashford RW, Thomson MC (2003) Risk mapping of visceral leishmaniasis: the role of local variation in rainfall and altitude on the presence and incidence of Kala-azar in Eastern Sudan. Am J Trop Med Hyg 68(1):10–17

Faucher B, Gaudart J, Faraut F, Pomares C, Mary C, Marty P, Piarroux R (2012) Heterogeneity of environments associated with transmission of visceral leishmaniasis in South-Eastern France and implication for control strategies. PLoS Negl Trop Dis 6(8):e1765. https://doi.org/10.1371/journal.pntd.0001765

Feliciangeli MD, Delgado O, Suarez B, Bravo A (2006) Leishmania and sand flies: proximity to woodland as a riskfactor for infection in a rural focus of visceral leishmaniasis inwest central Venezuela. Trop Med Int Health 11:1785–1791

Ferro CE, Cardenas D, Corredor A, Munstermann LE (1998) Life cycle and fecundity analysis of Lutzomyia shannoni (Dyar) (Diptera: Psychodidae). Mem Inst Oswaldo Cruz 93:195–199

Gage KL, Burkot TR, Eisen RJ, Hayes EB (2008) Climate and vector borne diseases. Am J Prev Med 35(5):436–450

Gálvez R, Descalzo MA, Miró G, Jiménez MI, Martín O, Dos Santos-Brandao F, Guerrero I, Cubero E, Molina R (2010) Seasonal trends and spatial relations between environmental/meteorological factors and leishmaniosis sand fly vector abundances in Central Spain. Acta Trop 115:95–102

Gebre-Michael T, Malone JB, Balkew M, Ali A, Berhe N, Hailu A, Herzi AA (2004) Mapping the potential distribution of Phlebotomus martini and P. orientalis (Diptera: Psychodidae), vectors of Kala-azar in East Africa by use of geographic information systems. Acta Tropica 90(1):73–86

Ghosh K, Mukhopadhyay J, Desai MM, Senroy S, Bhattacharya A (1999) Population ecology of phlebotomus argentipes (Diptera: Psychodidae) in West Bengal. India. J Med Entomol 36 (5):588–594

Gould P (1993) The slow plague: Geography of the AIDS pandemic. Cambridge: Blackwell Publishers

Gubler DJ (1991) Insects in disease transmission. Hunter Tropical Medicine. G. T.Strickland. Philadelphia, W. B. Saunders, pp 981–1000

Guerrant RL, Blackwood BL (1999) Threats to global health and survival: the growing crises of tropicalinfectious diseases—an "unfinished" agenda. Clin Infect Dis 28:966–986

Hansen AJ, Knight RL, Marzluff JM, Powell S, Brown K, Gude PH, Jones A (2005) Effects of exurban development on biodiversity: patterns, mechanisms, and research needs. Ecol Appl 15:1893–1905

Harvell CD, Mitchell CE, Ward JR, Altizer S, Dobson AP (2002) Climate warming and disease risks for terrestrial and marine biota. Sci 296:2158–2162

Hay SI, Battle KE, Pigott DM, Smith DL, Moves CL, Bhatt S, Brownstein JS, Collier N, Myers MF, George DB, Gething PW (2013) Global mapping of infectious disease. Philos Trans R SocLond B BolSci 368 (1614):20120250

Hay SI, Tucker CJ, Rogers DJ, Packer MJ (1996) Remotely sensed surrogates of meteorological data for the study of the distribution and abundance of arthropod vectors of disease. Ann Trop Med Parasitol 90:1–19

Herbreteau V, Salem G, Souris M, Hugot JP, Gonzalez JP (2005) Sizing up human health through remote sensing: uses and misuses. Parassitologia 47:63–79

Hermeto MV, Dias DV, Genaro O, Rotondo-Silva A, Costa CA, Toledo VPCP, Michalick MSM, Williams P, Mayrink W (1994) Outbreak of cutaneous leishmaniasis in the Rio Doce valley, Minas Gerais, Brazil. Mem Inst Oswaldo Cruz 89:519–521

Hongoh V, Hoen AG, Aenishaenslin C, Waaub JP, Bélanger D, Michel P (2011) The lyme-MCDA consortium spatially explicit multi-criteria decision analysis for managing vector-borne diseases. Int J Health Geogr 10:70. https://doi.org/10.1186/1476-072x-10-70

Høye TT, Forchhammer MC (2008) The influence of weather conditions on the activity of high-arctic arthropods inferred from long-term observations. BMC Ecol 8(1):8

Hudson PJ, Dobson AP, Lafferty KD (2006) Is a healthy ecosystem one that is richin parasites? Trends Ecol Evol 21:381–385

Joshi S, Bajracharya BL, Baral MR (2006) Kala-azar (visceral leishmaniasis) from Khotang. Kathmandu Univ Med J 4:232–234

Kaburi JC, Ngumbi PM, Anjili CO (2011) Sandfly-saliva injected during repeated feeding on a sensitized hamster causes fecundity and mortality to female Phlebotomus duboscqi (Diptera: Psychodidae). J Vect Borne Dis 48:61–63

Kalluri S, Gilrut P, Rogers D, Zzczur M (2007) Surveillence of arthroprod vector borne infectious disease using remote sensing techniques: a review. PLoS Pathog 3:1361–1371

Kamdem C, Fossog BT, Simard F, Etouna J, Ndo C, Kengne P et al (2012) Anthropogenic habitat disturbance and ecological divergence between incipient species of the malaria mosquito Anopheles gambiae. PLoS One 7(6):e39453

Karl TR (2003) Modern global climate change. Sci 302(5651):1719–1723

Kassem HA, Siri J, Kamal HA, Wilson ML (2012) Environmental factors underlying spatial patterns of sand flies (Diptera: Psychodidae) associated with leishmaniasis in southern Sinai. Egypt Acta Trop 123(1):8–15

Kesari S, Bhunia GS, Kumar V, Jeyaram, Ranjan A, Das P (2010) Study of house-level risk factors associated in the transmission of Indian Kala-azar. Parasit Vectors 3:94. https://doi.org/10.1186/1756-3305-3-94

Kesari S, Bhunia GS, Kumar V, Jeyaram A, Ranjan A, Das P (2011) A comparative evaluation of endemic and non-endemic region of visceral leishmaniasis (Kala-azar) in India with ground survey and space technology. Mem Inst Oswaldo Cruz, Rio de Janeiro, 106(5):515–523

Kesari S, Palit A, Kishore K (1992) Study of breeding habitats of sandflies preliminary approach. J Commun Dis 24:62–63

Killick-Kendrick R (1983) Investigation of Phlebotomine sandflies. In: Lumsden WHR, Evans DA Biology of the kinetoplastide, vol 2, London, Academic Press
Kim KS, Beresford RM, Henshall WR (2007) Prediction of disease risk using site-specific estimates of weather variables. New Zealand Plant Protect 60:128–132
Krishnaswamy J, Bawa KS, Ganeshaiah KN, Kiran MC (2009) Quantifying and mapping biodiversity and ecosystem services: utility of a multi-season NDVI based Mahalanobis distance surrogate. Remote Sens Environ 113(4):857–867
Kumar V, Kesari S, Dinesh DS, Tiwari AK, Kumar AJ, Kumar R, Singh VP, Das P (2015) A report on the indoor residual spraying (IRS) in the control of Phlebotomus argentipes, the vector of visceral leishmaniasis in Bihar (India): an initiative towards total elimination targeting 2015 (Series-1). J Vect Borne Dis 46:225–229
Kumar V, Kesari S, Kumar AJ, Dinesh DS, Ranjan A, Prasad M, Sinha NK, Kumar R, Das P (2009) Vector density and the control of Kala-azar in Bihar, India. MemInstOswaldo Cruz, Rio de Janeiro 104(7):1019–1022
Lacaux JP, Tourre YM, Vignolles C, Ndione JA, Lafaye M (2007) Classification of ponds from high-spatial resolution remote sensing: application to Rift Valley fever epidemics in Senegal. Remote Sens Environ 106:66–74
Leica Geosystems (2008) Leica geosystems geospatial imaging ERDAS IMAGINE® 9.2. Norcross, USA
Lewis DJ (1971) Phlebotomid sandflies. Bull World Health Org 44:535–551
Lewis DJ (1978) The phlebotomine sand flies (Diptera: Psychodidae) of the oriental region. Bull British Museum Nat Hist Entomol 37:217–343
Li ZL, Becker F (1993) Feasibility of land surface temperature and emissivity determination from AVHRR data. Remote Sens Environ 43:67–85
Lindgren E, Gustafson R (2001) Tick-borne encephalitis in Sweden and climate change. Lancet 358(9275):16–18
Lunetta RS, Elvidge CD (1998) Remote sensing change detection. MI: Ann Arbor Press. Magill AJ (1995) Epidemiology of the leishmaniases. Dermatol Clin 13(3):505–523
Malone JB, Huh OK, Fehler DP et al (1994) Temperature data from satellite imagery and the distribution of schistosomiasis in Egypt. Am J Trop Med Hyg 50:714–722
Malone JB, Yilma JM, McCarroll JC, Erko B, Mukaratirwa S, Zhou X (2001) Satellite climatology and the environmental risk of Schistosoma mansoni in Ethiopia and east Africa. Acta Trop 79:59–72
Markham BL, Storey JC, Williams DL, Irons JR (2004) Landsat sensor performance: history and current status. IEEE Trans Geosci Remote Sens 42(12):2691–2694
Martin LB, Hopkins WA, Mydlarz LD, Rohr JR (2010) The effects of anthropogenic global changes on immune functions and disease resistance. Acad Sci 1195:129–148
McMichael AJ (2004) Environmental and social influences on emerging infectious diseases: past, present and future. Phil Trans R Soc Lond B 359:1049–1058
McMichael AJ (2015) Extreme weather events and infectious disease outbreaks. Virulence 6 (6):543–7
McMichael AJ, Bouma MJ (2000) Global changes, invasive species and human health. In: Mooney HA, Hobbs RJ (eds) Invasive species in a changing world. Island Press, Washington DC, pp 191–210
Meade MS, Florin JW, Gesler WM (1988) Medical geography. The Guilford Press, New York
Moore JS, Kelly TB, Killick-Kendrick R, Killick-Kendrick M, Wallbanks R, Molyneux DH (1987) Honeydew sugars in wild-caught *Phlebotomusariasi* detected by high performance liquid chromatography (HPLC) and gas chromatography (GC). Med Vet Entomol 1:427–434
Morens DM, Folkers GK, Fauci AS (2004) The challenge of emerging andre-emerging infectious diseases. Nature 430:242–249
Morse SS (1995) Factors in the emergence of infectious diseases. Emerg Infect Dis 1(1):7–15
Morse SS, Schluederberg A (1990) Emerging viruses: the evolution of viruses and viral diseases. J Infect Dis 162:1–7

Moser W, Greter H, Schindler C, Allan F, Ngandolo BN, Moto DD, Utzinger J, Zinsstag J (2014) The spatial and seasonal distribution of Bulinustruncatus, Bulinusforskalii and Biomphalariapfeifferi, the intermediate host snails of schistosomiasis, in N'Djamena. Chad Geospat Health 9(1):109–118

Mukhopadhyay AK, Rahman SJ, Chakravarty AK (1990) Effects of flood control on immature stages of sandflies in flood prone Kala-azar endemic villages of North Bihar, India. WHO/VBC.90:986

Müller GC, Schlein Y (2006) Sugar questing mosquitoes in arid areas gather on scarce blossoms that can be used for control. Int J Parasitol 36:1077–1080

Murdock CC, Paaijmans KP, Cox-Foster D, Read AF, Thomas MB (2012) Rethinking vector immunology: the role of environmental temperature in shaping resistance. Nature Rev Microbiol 10:869–876

Napier LE (1926) An epidemiological consideration of the transmission of Kala-azar in India. India Med Res Memoir 4:219–265

Napier LE (1962) An epidemiological consideration of the transmission of Kala-azar in India. India Med Res Memoir. 4:219–265

Napier LE, Smith ROA (1926) A study of the bionomics of Phlebotomusargentipes, with special reference to the conditions in Calcutta. Indian Med Res Mem 4:161–172

Nieto P, Malone JB, Bavia ME (2006) Ecological niche modeling for visceral leishmaniasis in the state of Bahia, Brazil, using genetic algorithm for rule-set prediction and growing degree day-water budget analysis. Geospatial Health 1:115–126

Nizeyi JB, Mwebe R, Nanteza A, Cranfeld MR, Kalema GRNN, Graczyk TK (1999) Cryptosporidium sp. and Giardia sp. Infections in Mountain Gorillas (Gorilla gorilla beringei) of the Bwindi Impenetrable National Park, Uganda. J Parasitol 85(6):1084–1088

Nurminen M, Nurminen T, Corvalan CF (1999) Methodologic issues in epidemiologic risk assessment. Epidemiology 10(5):585–593

Odiit M, Bessell PR, Fèvre EM, Robinson T, Kinoti J, Coleman PG, Welburn SC, McDermott J, Woolhouse MEJ (2006) Using remote sensing and geographic information systems to identify villages at high risk for rhodesiense sleeping sickness in Uganda. Trans R Soc Trop Med Hyg 100:354–362

Ostyn B, Vanlerberghe V, Picado A, Dinesh DS, Sundar S, Chappuis F, Rijal S, Dujardin JC, Coosemans M, Boelaert M, Davies C (2008) Vector control by insecticide-treated nets in the fight against visceral leishmaniasis in the Indian subcontinent, what is the evidence? Trop Med Int Health. 13(8):1073–1085

Ozer N (2005) Emerging vector-borne diseases in a changing environment. Turkish J Biol 29:125–135

Patz J, Confalonieri UEC (2005) Ecosystem regulation of infectious diseases Conditions and Trends. Millenn Ecosyst Assess, Island Press, Washington, pp 391–415

Patz J, Wolfe N (2002) Global ecological change and human health. In: Conservation medicine: ecological health in practice (ed) Aguirre A, Ostfeld R, Tabor G, House C, Pearl M), pp 167–181. Oxford University Press

Patz JA, Graczyk TK, Geller N, Vittor AY (2000) Effects of environmental change on emerging parasitic diseases. Int J Parasitol 30:1395–1405

Pavlovsky EN (1966) The natural nidality of transmissible disease (N.D. Levine, ed.). University of Illinois Press, Urbana

Peterson AT, Martinez-Campos C, Nakazawa Y, Martinez-Meyer E (2005) Time-specific ecological niche modeling predicts spatial dynamics of vector insects and human dengue cases. Trans R Soc Trop Med Hyg 99(9):647–655

Pfeiffer DU, Robinson TP, Stevenson M, Stevens KB, Rogers DJ, Clements ACA (2008) Spatial analysis in epidemiology. Oxford University Press, Oxford, UK

Picado A, Das ML, Kumar V, Dinesh DS, Rijal S et al (2010a) Phlebotomus argentipes seasonal patterns in India and Nepal. J Med Entomol 47(2):283–6

Picado A, Murari LD, Vijay K, Diwakar SD, Suman R, Shri PS, Pradeep D, Marc C, Marleen B, Clive D (2010b) Phlebotomusargentipes seasonal patterns in India and Nepal. J Med Entomol 47(2):283–286

Picado A, Singh SP, Rijal S, Sundar S, Ostyn B, Chappuis F, Uranw S, Gidwani K, Khanal B, Rai M, Paudel IS, Das ML, Kumar R, Srivastava P, Dujardin JC, Vanlerberghe V, Andersen AW, Davies CR, Boelaert M (2010) Longlasting insecticidal nets for prevention of Leishmania donovani infection in India and Nepal: paired cluster randomised trial. British Med J 341. https://doi.org/10.1136/bmj.c6760

Povoa M, Wirtz R, Lacerda R, Miles M, Warhurst D (2001) Malaria vectors in the municipality of Serra do Navio, State of Amapa, Amazon Region. Brazil Mem Inst Oswaldo Cruz 96:179–184

Qiu ZY, Li J, Guo HJ (1998) Application of remote sensing technique. Wuhan University Press, Wuhan, China, pp 97–98

Quintana MG, Fernández MS, Salomón OD (2012) Distribution and abundance of phlebotomine, vectors of Leishmaniasis, in Argentina: spatial and temporal analysis at different scales. J Tropical Med. https://doi.org/10.1155/2012/652803

Rahman A (2008) Climate change and its impact on health in Bangladesh. Reg Health Forum 12:16–26. http://www.searo.who.int/LinkFiles/Regional_Health_Forum_Volume_12_No_1_Climate_change_and_its_impact.pdf

Rahman KM, Islam S, Rahman MW et al (2010) Increasing incidence of post-Kala-azar dermal leishmaniasis in a population-based study in Bangladesh. Clin Infect Dis 50(1):73–76

Raina S, Mahesh DM, Kaul R, Satinder KS, Gupta D, Sharma A, Thakur S (2009) A new focus of visceral leishmaniasis in the Himalayas. India. J Vect Borne Dis 46:303–306

Rakotomanana F, Randremanana RV, Rabarijaona LP, Duchemin JB, Ratovonjato J, Ariey F, Rudant JP, Jeanne I (2007) Determining areas that require indoor insecticide spraying using Multi Criteria Evaluation, a decision-support tool for malaria vector control programmes in the Central Highlands of Madagascar. Int J Health Geogr 6:2. https://doi.org/10.1186/1476-072x-6-2

Ramasamy R, Surendran SN (2016) Mosquito vectors developing in atypical anthropogenic habitats: global overview of recent observations, mechanisms and impact on disease transmission. J Vect Borne Dis 53:91–98

Ranjan A, Sur D, Singh VP, Siddique NA, Manna B, Lal CS, Sinha PK, Kishore K, Bhattacharya SK (2005) Risk factors for Indian Kala-azar. Am J Trop Med Hyg 73(1):74–78

Raso G (2006) An integrated approach for risk profiling and spatial prediction of Schistosoma mansoni—hookworm coinfection. In: Proceedings of the National Academy of Sciences of the USA, 103:6934–6939

Raulerson JT (2010) Singularities: technoculture, transhumanism, and science fiction in the 21st Century. PhD (Doctor of Philosophy) thesis, University of Iowa. http://ir.uiowa.edu/etd/2968

Ready PD (2008) Leishmaniasis emergence and climate change. Rev Sci Tech Off Int Epiz 27 (2):399–412

Rejmankova E, Savage PM, Rejmanek M, Arredondo JI, Roberts DR (1991) Multivariate analysis of relationships between habitats, environmental factors and occurence of Anopheline mosquito larvae Anopheles albimanus and A. *pseudopunctipennis* in Southern Chiapas. Mexico J Appl Ecol 28:827–841

Remaudière G (1992) A simplified method for mounting aphids and other small insects in Canada balsam. Rev Front Entomol 14:185–186

Richards JA (1986) Remote sensing digital image analysis. Springer, Berlin

Rodo X, Pascual M, Fuchs G, Faruque AS (2002) ENSO and cholera: a non-stationary link related to climate change? In: Proceedings of the National Academy of Sciences (USA), 99:12901–12906

Rogers DI, Hay SI, Packer ML (1996) Predicting the distribution of tsetse flies in West Africa using temporal Fourier processed meteorological satellite data. Ann Tropical Med Parasitol 90:225–241

Rogers DJ, Randolph SE, Snow RW, Hay SI (2002) Satellite imagery in the study and forecast of malaria. Nat 415(6872):710–715

Rosenfield GH, Fitzpatrick L (1986) Kappa coefficient of agreement as a measure of thematic classification accuracy. Photogram Eng Remote Sens 52:223–227

Roujean JL, Breon FM (1995) Estimating PAR absorbed by vegetation from bidirectional reflectance measurements. Remote Sens Environ 51(3):375–384. https://doi.org/10.1016/0034-4257(94)00114-3

Saaty TL (1980) The analytic hierarchy process, New York: McGraw Hill. International, Translated to Russian, Portuguese, and Chinese, Revised editions, Paperback (1996, 2000), Pittsburgh: RWS Publications

Saaty TL (2005) The analytic hierarchy and analytic network process for the measurement of intangible criteria and for decision making. In: Figueira J, Salvatore G, Ehrgott M (eds) Multiple criteria decision analysis: state of the art surveys. Springer, Berlin Heidelberg: New York, NY, USA, pp 345–407

Schlein Y, Jacobson RL (1999) Sugar meals and longevity of the sandfly Phlebotomus papatasi in an arid focus of Leishmania major in the Jordan Valley. Med Vet Entomol 13(1):65–71

Schlein Y, Warburg A (1986) Phytophagy and the feeding cycle of *Phlebotomuspapatasi* (Diptera: Psychodidae) under experimental conditions. J Med Entomol 23:11–15

Schmidt K, Ostfeld R (2001) Biodiversity and the dilution effect in disease ecology. Ecology 82:609–619

Sharma VP (1995) Research on newer strategies for vector control. South-East Asia advisory committee on health research twenty-first session. SEA/ACHR/21/8, pp 10–13 April 1995. http://www.searo.who.int/LinkFiles/Technical_Documents_achr-21-8.pdf

Sharma U, Singh S (2008) Insect vectors of Leishmania: distribution, physiology and their control. J Vect Borne Dis 45:255–272

Silva F, Gomes R, Prates D, Miranda JC, Andrade B, Barral-Netto M, Barral A (2005) Inflammatory cell infiltration and high antibody production in BAL/c mice caused by natural exposure to Lutzomiya longipalpis bites. Am J Trop Med Hyg 72:94–98

Singh SM (1988) Brightness temperature algorithms for landsat thematic mapper data. Remote Sens Environ 24:509–512

Singh R, Lal S, Saxena VK (2008a) Breeding ecology of visceral leishmaniasis vector sandfly in Bihar state of India. Acta Trop 107:117–120

Singh A, Roy SP, Kumar R, Nath A (2008b) Temperature and humidity play a crucial role in the development of *P. argentipes*. Journal of Ecophysiology& Occupational Health 8(1 & 2)

Singh BB, Sharma R, Gill JPS, Aulakh RS, Banga HS (2011) Climate change, zoonoses and India. Rev Sci Tech Off Int Epiz 30(3):779–788

Sivagnaname N, Amalraj DD (1997) Breeding habitats of vector sandflies and their control in India. J Commun Dis 29:153–159

Smith ROA (1959) Bionomics of P. *argentipes*. Bulletine of Calcutta School Trop Med 7: 19–21

Sobrino JA, Jimenez-Munoza Paolini (2004) Land surface temperature retrieval from LANDSAT—TM 5. Remote Sens Environ 92:521–534

Soti V, Tran A, Bailly J-S, Puech C, Lo Seen D, Bégué A (2009) Assessing optical earth observation systems for mapping and monitoring temporary ponds in arid areas. Int J Appl Earth Obs Geoinf 11:344–351

Stevens KB, Pfeiffer DU (2011) Spatial modelling of disease usingdata- and knowledge-driven approaches. Spat Spatio Temporal Epidemiol. 2:125–133

Sudhakar S, Srinivas T, Palit A, Kar SK, Bhattacharya SK (2006) Mapping of risk prone areas of Kala-azar (Visceral leishmaniasis) in parts of Bihar state, India: an RS and GIS approach. J Vect Borne Dis 43:115–122

Thakur CP (2007) A new strategy for elimination of Kala-azar from rural Bihar. Indian J Med Res 126:447–451

Thanyapraneedkul J, Muramatsu K, Daigo M (2009) Improvement accuracy of terrestrial NPP estimation using ADEOS-II/GLI data, pp 78–88. Available at: https://doors.doshisha.ac.jp/duar/repository/ir/18059/038010020007.pdf

Theodor O (1936) On relationship of Phlebotomus papatasi to the temperature and humidity of the environment. Bull Entomol Res 27:653–671

Thomson M, Indeje M, Connor S, Dilley M, Ward N (2003) Malaria early warning in Kenya and seasonal climate forecasts. The Lancet 362:580
Toumi A, Chlif S, Bettaieb J, Alaya NB, Boukthir A, Ahmadi ZE, Salah AB. (2012) Temporal Dynamics and Impact of Climate Factors on the Incidence of Zoonotic Cutaneous Leishmaniasis in Central Tunisia. PLoS Negl Trop Dis 6(5):e1633
Tsiko RG, Haile TS (2011) Integrating geographical information systems, fuzzy logic and analytical hierarchy process in modelling optimum sites for locating water reservoirs: a case study of the Debub District in Eritrea. Water 3:254–290. https://doi.org/10.3390/w3010254
Tucker CJ (1979) Red and photographic infrared linear combinations for monitoring vegetation. Remote Sens Environ 8(2):127–150
Tucker CJ, Justice CO, Prince SD (1986) Monitoring the grasslands of the Sahel 1983-1985. Int J Remote Sens 7:1571–1582
Valenzuela GJ, Belkaid Y, Garfeild MK, Mendez S, Kamhawi S, Rowton ED, Sacks DL, Libeiro JM (2001) Toward a defined anti-leishmania vaccine targeting vector antigens: characterization of a protective salivary protein. J Exp Med 194:331–342
Valor E, Caselles V (1996) Mapping land surface emissivity from NDVI: applications to European, African, and South American areas. Remote Sens Environ 57:167–184
Van De Griend AA, Owe M (1993) On the relationship between thermal emissivity and the normalized difference vegetation index for natural surfaces. Int J Remote Sensing 14(6):1119–1131
Victora CG, Huttly SR, Fuchs SC, Olinto MT (1997) The role of conceptual frameworks in epidemiological analysis: a hierarchical approach. Int J Epidemiol 26:224–227
Wallis J, Lee K (1999) Primate conservation: the prevention of disease transmission. Int J Primatol 20:803–826
Waring S, Zakos-Feliberti A, Wood R, Stone M, Padgett P, Arafat R (2005) The utility of geographic information systems (GIS) in rapid epidemiological assessments following weather-related disasters: methodological issues based on the tropical storm allison experience. Int J Hyg Environ Health 208:109–116
Werneck GL (2008) Forum: geographic spread and urbanization of visceral leishmaniasis in Brazil. Introduct Cad Saúde Pública 24(12):2937–2940
Williams J (1997) The weather book, Vintage Books, ISBN 0–679-77, pp 665–666
Woolhouse ME, Gowtage-Sequeria S (2005) Host range and emerging and reemerging pathogens. Emerg Infect Dis 11:1842–1847
World Health Organization (1995) Communicable disease prevention and control: new, emerging, and re-emerging infectious diseases. WHO Doc. A48/15; Feb. 22
World Health Organization (WHO) (2007) Guidelines and standard operating procedure for Kala-azar elimination in South-East Asia countries. WHO (Trial Edition) 19 Nov 2007. http://www.searo.who.int/LinkFiles/Kala_azar_VBC-85_Rev_1.pdf
World Health Organization (2009) Global health risks: mortality and burden of disease attributable to selected major risks. ISBN 978 92 4 156387 1. http://www.who.int/healthinfo/global_burden_disease/GlobalHealthRisks_report_full.pdf
World Health Report (1996) Fighting disease, fostering development, p 15. Geneva: WHO. http://www.who.int/whr2001/2001/archives/1996/index.htm
Wu WP, Davis G, Liu HY, Seto E, Lu SB, Zhang J, Hua ZH, Guo JG, Lin DD, Chen HG, Peng G, Feng Z (2002) Application of remote sensing for surveillance of snail habitats in Poyang Lake, China. Chin J Parasitol Parasit Dis 20(4):205–208
Yi Y, Yang D, Chen D, Huang J (2007) Retrieving crop physiological parameters and assessing water deficiency using MODIS data during the winter wheat growing period. Can J Remote Sens 33(3):189–202
Ying X, Zeng GM, Chen GQ, Tang L, Wang KL, Huang DY (2007) Combining AHP with GIS in synthetic evaluation of eco-environment quality—a case study of hunan province. China Ecol Model 209(2–4):97–109

Zeilhofer P, dos Santos ES, Ribeiro ALM, Miyazaki RD, dos Santos MA (2007) Habitat suitability mapping of Anopheles darlingi in the surroundings of the Manso hydropower plant reservoir, MatoGrosso, Central Brazil. International Journal of Health Geographics 6:7. https://doi.org/10.1186/1476-072x-6-7

Zilberstein D, Shapira M (1994) The role of pH and temperature in the development of Leishmania parasites. Annu Rev Microbiol 48:449–470

Chapter 6
Disease Risk Assessment and GIS Technology

6.1 Introduction

Disease risk analysis includes scientifically assembling data and undertaking risk calculations of predefined menace and susceptibility. This is assumed in the planning phase of an interposition. An early warning system (EWS) is a sequence of information communication systems (ICS), providing time for the retort structure to concoct for the hostile result and to abate its impact (Santos-Reyes and Beard 2013). EWS is the key component of disease risk lessening, thus empowering activities to be taken to alleviate latent damage, and sometimes, providing an opportunity to prevent the hazardous event from stirring (Bouchard et al. 2013). The aim of a disease risk and EWS would be to offer public health administrators and the common civic with as much advance sign as potential about the probability of a disease epidemic in a certain situation, thus spreading the variety of practicable response decisions.

To accomplish an effective menace lessen in grole, EWS should be tacit as an information system intended to assist decision building of the pertinent nationwide and regional level organizations (Istudor et al. 2016). The system also permit susceptible persons and social clutches to yield actions to alleviate the influences of an imminent risk. In contrast, environmental observations and weather predictions can hypothetically be engaged in efforts to forecast the advent of a host-pathogen relationship and so permit prospects to minimalize its propagation.

6.2 Components of Early Warning System

Epidemiological surveillance: Epidemiological reconnaissance systems arrange for timely diffusion of information. Traditionally, surveillance has concentrated explicitly on observing the incidence of disease. However, surveillance also

G. S. Bhunia and P. K. Shit, *Geospatial Analysis of Public Health*,
https://doi.org/10.1007/978-3-030-01680-7_6

prerequisites to comprise observing of vicissitudes in vector population profusion. These kinds of ecologically-based surveillance systems are commonly recognized in some circumstances, for example, in epidemiological surveillance for vector borne disease (Thompson and Etter 2015; Shah and Gupta 2013; Ozair et al. 2012), West Nile fever (Swaddle and Calos 2008), Rift Valley fever (Métras et al. 2011), Leishmaniasis (Medenica et al. 2015), and Japanese encephalitis (Wang et al. 2014).

Ecological annotations. Systematic observations of environmental data is a significant constituent of EWS. Such as, whether a substantial rainfall incident will tip to flooding may reliant on the current antiquity of precipitation in the area (Rajabi 2015). Conversely, satellite based information for monitoring ecological parameters such as land use/land cover, soil moisture, vegetative cover, and sea surface temperature on a global/regional scale will unavoidably play an imperative part in vector-borne and/or infectious disease EWS (Kesari et al. 2013). Wang et al. (2014) studied epidemiological features and identify high relative risk space-time Hand, foot, and mouth disease clusters at a fine spatial scale.

Susceptibility assessment: Susceptibility assessment is closely associated with the public health surveillance. Vulnerability denotes to a population's compassion to a menace, along with its capability to deal with the hazard. Sutherst (2004) reported that climate change is worldwide in nature and poses indefinite future menaces to anthropogenic and natural environments, other local variations are stirring more quickly on a global measure and are having important possessions on vector-borne diseases (VBDs). Consequently, the ecology and epidemiology of VBDs can be defined using the "disease triangle" of host-pathogen-environment initially established by plant pathologists (Sutherst 2000). Musa et al. (2013) used of GIS technology as a public health tool for Cholera to Cancer.

Risk analysis: Risk analysis is conducted to allocate detailed prospects to the likely influences of an imminent hazard (Frössling et al. 2012). The risk calculation may be depend on professional estimation, the occurrence of identified risk factors, and/or the yields of qualitative or quantitative risk analyses (Reist et al. 2012). In developing countries, risk patterns are quickly varying as a consequence of climate change, environmental modification, urbanization, population growth, migration, economic change and environmental degradation etc. So far risk analysis is still deeply immobile in character in numerous spaces due to absence of up-to-date cartographic and census information to demarcate actual risk levels (Bobadilla Suarez et al. 2017).

Awareness/response. Warning schemes must be established in recital with progresses in local, regional, or national level response capabilities, mainly in highly endemic areas. The activities taken to alleviate disease outbursts can often transmit momentous expenditures and can pretense a huge encumbrance on assets in developing countries (Bayer et al. 2009). Furthermore, execution of control measures (such as insecticides spraying, vector control, public awareness) can occasionally look active public confrontation. Such possible consequences and cost-benefit contemplation essential to be prudently evaluated with the intention of response strategies can be adjusted to greatest ensemble the priorities, requirements, and capability of the local public.

Public communication. To confirm that the cautioning information and suggested response approaches are noted by the populations at jeopardy, effective public communication plans must be established (Medford-Davis and Kapur 2014). If the warning consultant dearth of status in the area concerned or has an antiquity of inconsistent interactions with local groups, cautions may not be well acknowledged. Similarly, if lots of warnings had been assumed earlier or no caution was provided at all during the earlier disease outbreaks, the reliability of imminent warnings will be jeopardized. Finally, disease early warnings must embrace clear elucidations of the real level of risk tangled and of the specific clutches that are most susceptible.

6.3 Role of Earth Observation in Disease Risk Analysis and Early Warning System

Earth observation (EO), through measuring and monitoring, provides an insight and understanding into Earth's complex processes and changes. EO includes measurements that can be made directly or by sensors in situ or remotely to provide key information to models or other tools to support decision making processes (Hu et al. 2014). EO assists governments and civil society to identify and shape corrective and new measures to achieve sustainable development through original, scientifically valid assessments and early warning information on the recent and potential long-term consequences of human activities on the biosphere (Eisen and Eisen 2011). At a time when the world community is striving to identify the impacts of human actions on the planet's life support system, time sequenced satellite images help to determine these impacts and provide unique, visible and scientifically convincing evidence that human actions are causing substantial changes to the Earth's environment and natural resource base (i.e. ecosystems changes, urban growth, transboundary pollutants, loss of wetlands, etc.). The usage of geospatial models comprise risk index analysis for forecasting encephalitis virus transmission (Van der Kelen 2014) and human WNV infection (Kilpatrick and Pape 2013), and WNV vector abundance in Greater Toronto (Yoo 2014), and near real-time monitoring of land surface temperatures and the re-normalized difference vegetation index (NDVI) to predict VL (Bhunia et al. 2012a, b).

By enhancing the visualization of scientific information on environmental change, satellite imagery will enhance environmental management and raise the awareness of emerging environmental threats (Bhunia et al. 2012a, b). EO provides the opportunity to explore, to discover, and to understand the world in which we live from the unique vantage point of space.

6.4 Spatial Scale of Early Warning System

International level: Cyclical climate predictions can be established and/or circulated by administrations such as the International Research Institute for Climate Prediction, NOAA's Climate Prediction Center, the European Centre for Medium-Range Weather Forecasting, and the World Meteorological Organization (WMO). Technical support and training can be provided by National Institutes of Health (NIH), National Public Health Institutes (NPHIs), World Health Organization (WHO), the Pan American Health Organization (PAHO), International Association of National Public Health Institutes (IANPHI), Vector and Vector-Borne Diseases Research Institute (VVBDI) to aidimprove national competences for investigation and EWS.

National level: Several Scientific/Research teams employed within national public health agencies (Indian Council of Medical Research) can usage climate prognostications together with evidence from susceptibility determinants and surveillance systems to evaluate the disease menaces modelled to definite groups. Consequently, national administrations and scientific agencies can also support strengthen the elementary setup for epidemiological and ecological reconnaissance structures. For example, National Vector Borne Diseases Control Programme (NVBDCP) is involved with technical experts in the field of public health, toxicology entomology, epidemiology and parasitology in different aspects of vector borne disease in India.

Regional level: Regional level EWS has been ascribed to better monitoring and predicting of epidemic phenomenon by investigating risks and integration of risk information in alternative planning and warnings at regional scale. Regional Institutions and Organizations play a role in providing specialized knowledge and advice which supports national efforts to develop and sustain early warning capabilities. For example, Rajendra Memorial Research Institute of Medical Science (RMRIMS) has been involved in regional and state level analysis of Visceral Leishmaniasis (Kala-Azar) elimination programme. National Institute of Malaria Research (NIMR) involved in short term as well as long-term elucidations to the hitches of malaria through basic, applied and operational field research in India.

Local level: Like communities and individuals can aid in activities such as monitoring vulnerability patterns, assessing intervention strategies and local response capacity, and strengthening networks for decision making and public communication.

6.5 Case Study 1: Assessment of Visceral Leishmaniasis Risk in Muzaffarpur District (Bihar), India: A GIS Approach

6.5.1 Introduction

Visceral leishmaniasis (VL), locally known as kala-Azar in India is a vector borne parasitic disease caused by *Leishmania donovani* that has become a major international public health concern (Thakur 2007). The disease is transmitted by the bite of infected female sandfly of the genus Phlebotomus *(Phlebotomus argentipes*) (Thakur 2007). The illness is predominantly found in tropical and sub-tropical climate regions around the world (WHO 1998), mostly common in Bangladesh, Brazil, India, Nepal and Sudan (Ryan and Ray 2004). The incidence of VL has grown dramatically around the world in recent decades (Guerin et al. 2002; Assimina et al. 2008). Undoubtedly, VL is a potential threat for the Indian sub-continent and one of the most neglected diseases in developing countries (Sudhakar et al. 2006). In the past, various control measures carried out in this area; however, failed to permanently decrease the number of cases (Dhiman and Yadav 2016; Kumar et al. 2015; WHO 2015).

The mapping of disease can be used to locate the areas where outbreaks originate and effectively target high-risk areas for early prevention control. Geographic information systems (GIS) are currently recognized as a set of strategic and analytic tools for public health (Riner et al. 2004; Bhunia et al. 2013), so the design and implementation of an information system for VL control with GIS capacity should be considered. Previous research on VL incidence has primarily focused on the association of the disease vis-à-vis vector with the environmental variables (Bhunia et al. 2010a, b, 2012a, b).

From, VL cases vary from one place to another; the spatial and time components must also be taken into consideration. These spatial distributions, or patterns, are of interest to many areas of geographic research because they can help identify and quantify patterns of features in space so that the underlying cause of the distribution can be determined (Fotheringham et al. 2002; Srividya et al. 2002). Geo statistical procedures explicitly describe the spatial relationships between endemic villages (Bhunia et al. 2013). Moreover, the identification of clusters via geo-statistical analyses, and geo-referenced epidemiological incidence report can be useful for disease monitoring and control programme. The investigation aids to epidemiology and medical entomology, with one of the aims being to resolve whether the clustering is statistically significant and admirable of further investigation, or whether it is likely to be a chance occurrence. Global and local indicators of spatial association like *Moran's I* (Cliff and Hord 1981) or Getis-Ord Statistics (Getis and Ord 1992) are often used to measure the data clustering level.

Therefore, the present study has been conducted to delineate the temporal and spatial analyses of VL using GIS techniques to understand the trend and dynamics of transmission, to identify risk areas, and to determine environmental factors associated with increased risk of kala-Azar.

6.5.2 Materials and Methods

Area of study

The Muzaffarpur District is located between 25° 54′–26° 23′ N latitude and 84° 53′–85° 45′ E longitude. This district is one of the oldest and largest trade centers in the entire state. It shares boundaries with East Champaran, Sitamarhi, Vaishali, Saran, Darbhanga, Samastipur and Gopalganj district covering an area of 3,172 km^2 having total population is 3 million. The region is plain and flat topography, located on the middle Gangetic plain, and contains some streams of water running from north to south. The climate is subtropical, being hot during summer with a range of temperature between 35° and 40 °C, while in winter it is cold with a mean temperature of 21 °C in January. Relative humidity (RH) is relatively high and usually between 50 and 70%, sometimes reaching 90%. Maximum rain receives during the monsoon, with an average annual rainfall of 165 mm.

Acquisition of VL incidence report and database preparation

A longitudinal study was carried out to understand the mechanism of disease transmission in the VL endemic region. The data on monthly VL incidence from 2005 to 2011 (7 years) were obtained from Public Health Centers (PHCs) and Bihar State Health Society (India). The number of VL cases per year was calculated for each village. The annual incidence rate was obtained by taking into account the number of inhabitants within each village (data obtained from Public Health Centers and State Health Society of Bihar, India) per 10,000 population. For spatial analysis, the variable retained for each village was the average of annual incidence rates observed during the whole study period (2005–2011). A digital database was generated in VEKPRO (Vector program) in relation to the location of the VL affected villages. In order to synchronize the digital data coordinate system, we used the Universal Transverse Mercator (UTM) projection system and World Geodetic System (WGS 84) datum. The coordinate VL incidence villages were recorded to measure the spatial correlation of VL. The location of each affected villages was mapped based on the assumption that most people become infected with VL near their neighboring villages.

Sampling of adults Sandfly

During the 2005–2011, the adult sand flies were collected once a month from the human dwellings and cattle sheds using ten light traps in ten highly endemic villages. The light traps were run from 6.00 p.m. to 6.00 a.m. in each sampling station (Njabo et al. 2009). Three types of sand flies species were collected, like *Phlebotomus argentipes, Phlebotomus Paptasi* and *Sergentomiya.* In this study, only *Phlebotomus argentipes* was considered as it is the proven vector of Indian VL. The counts of *Phlebotomus argentipes* in the traps were used to calculate mean monthly numbers of this adult sand flies/trap-night, as a measure of the density of the local vector population.

Meteorological variables

Weather data, like temperature (maximum and minimum), humidity (morning and evening), rainfall, and mean wind velocity (MWS) were obtained for 2005–2011 from the nearest weather station (Agricultural Research Institute, Pusa (Samastipur), Bihar), placed at about 25 km of distance from the Muzaffarpur district. Monthly averages for temperature, humidity, rainfall and MWS were calculated.

Spatial and geo-statistical analysis

In this study, the Cluster and Outlier (CO type) analysis was performed to identify clusters of VL incidence villages from 2005 to 2011 using Anselin's Local Moran's *I* statistic via Cluster and Outliers Analysis tool in ESRI's ArcGIS 9.3 Spatial Statistics toolbox (Mitchell 2005). The CO type analysis was executed based on Fixed Distance Band method, where each feature was analyzed within the context of those neighboring villages within 1 km distance. Villages outside the critical distance (<1 km) of a target feature do not influence calculations for that village. The CO type discriminates between a statistically significant ($P < 0.05$) cluster of high values (HH), cluster of low values (LL), outlier in which a high value is enclosed by low values (HL), and outlier in which a low value is bounded by high values (LH). The Z score and p-value represent the statistical significance of the computed index value (Mitchell 2005; Anselin 1995).

Subsequently, hotspot and coldspot analysis of VL incidences villages of Muzaffarpur district was analused using *Getis-Ord Gi** statistic to determine more intense the clustering of high VL incidence villages (hot spot) and the more intense the clustering of low VL incidence villages (cold spot). The *Gi** statistic returned for each feature in the dataset is a Z score that helps to decide whether or not to reject the null hypothesis. The p-value is the probability and has deceptively excluded the null hypothesis. For statistically significant positive Z scores, the larger the Z score is, the more intense the clustering of high VL incidence village and the significant negative Z scores (the smaller the Z score) is the more intense the clustering of low VL incidence village (Mitchell 2005).

Lastly, potential VL risk areas were delineated using probability ordinary kriging interpolation technique (Bhunia et al. 2012a, b). Ordinary kriging estimation is an effective tool to identify high-risk areas within polygon patterns of disease incidence by producing a smooth, continuous surface that defines the level of risk for that area (Gething et al. 2006; Croner and De Cola 2001; Kleinschmidt et al. 2000). The Root Mean Square Error (RMSE) error is calculated to measure the accuracy of the model.

Statistical analysis

Data was analysed using the statistical software SPSS ver. 10.0. Pearson's correlation coefficient was used to estimate the relationship between meteorological variables and *P. argentipes* abundance and disease incidence at the confidence level of 5%. The t-test was used to verify the differences in sand fly abundance and meteorological variables among seven consecutive years.

6.5.3 Results

The number of notifying VL cases in Muzaffarpur district during the period from 2005–2011 are presented in the Fig. 6.1. The number of VL cases during that period was 3,251 cases in 2005, fluctuated reaching a maximum of 4,920 cases in 2007, and declined reaching 2,531 cases in 2011 (Fig. 6.2). However, the maximum number of cases was recorded from the Paroo PHC (2571), whereas, lowest number of cases were recorded from Dholi PHC (260) during the study period. The proportion of cases reported to by the PHCs varied over the study period. The VL incidence rate was higher in 2007 (11.41/10,000 population, while cause-specific mortality rate was also high (4.1%) in this year. However, the case fatality rate was maximum (0.95) in 2008 in the district, during the study period. Table 6.1 shows the details epidemiological characteristics of Muzaffarpur district from 2005 to 2011. The incidence rate for each year was calculated separately and the average value of the disease incidence rate was used for further analysis. Conversely, it is found that the lowest incidence rate (5.15/10,000 population) recorded in 2009, and the case fatality rate calculated as 0.47%, followed by cause-specific mortality rate as 1.1%. The mean number of reported VL cases per year during the period is calculated as 3347.29 (SD $\pm$ 959.05).

Sand fly density

During the period between January 2005 and December 2011, a total of 1647 sand flies were collected, which consisted of 39.86% males and 60.14% females. Figure 6.3 represents the monthly distribution of sand fly density during the period from 2005 to 2011. The highest sand fly density was recorded in the month of September (10.72 $\pm$ 1.76 trap per-night), whereas, lowest density was evidenced in January (0.59 $\pm$ 0.45 trap per night).

Climatic characteristics

During the study period (2005–2011), the average maximum temperature of this region was 30.43 °C; whereas, the average minimum temperature was 19.65 °C. The highest maximum temperature was recorded in April, 2010 (38.3 °C), whereas, the lowest minimum temperature was recorded as 6.9 °C during the study period. However, the analysis of air temperature divulges that December and January are the coldest months of the year, while April and May are the hottest. The annual

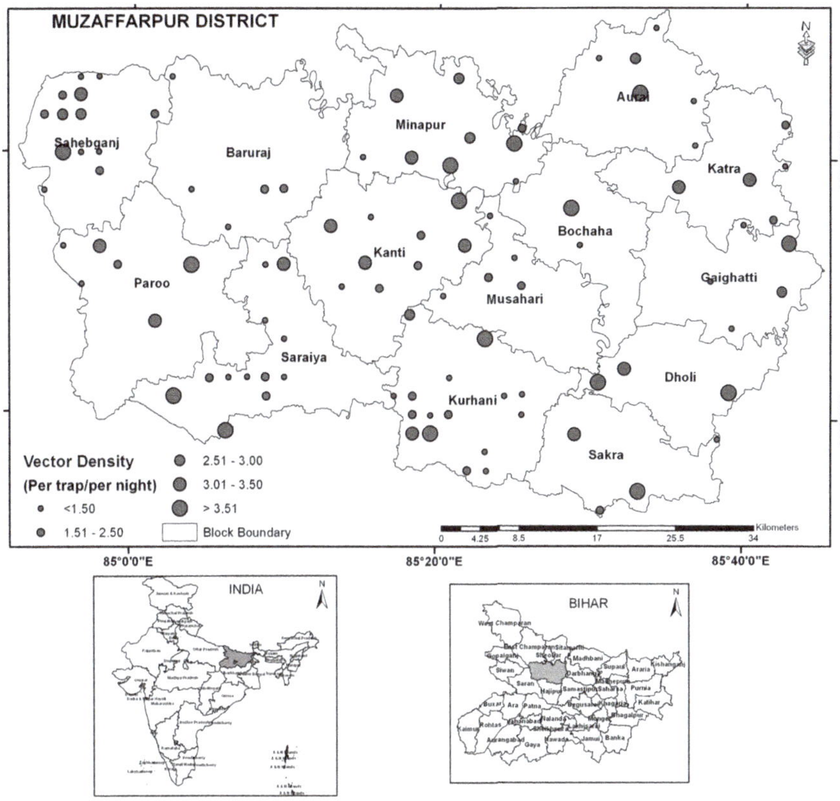

Fig. 6.1 Location map of the study area and vector density

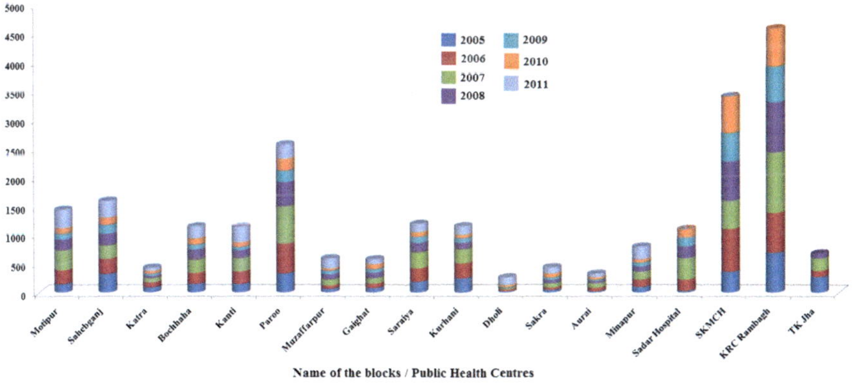

Fig. 6.2 Block-wise Kala-Azar distribution in Muzaffarpur district during the period between 2005 and 2011

Table 6.1 Epidemiological characteristics of VL incidence in Muzaffarpur district from 2005 to 2011

Year	Number of cases	Incidence rate per 10,000 population	Case fatality rate (%)	Cause-specific mortality rate (%)
2005	3251	7.90	0.89	2.9
2006	4128	9.80	0.90	3.7
2007	4920	11.41	0.83	4.1
2008	3699	8.38	0.95	3.5
2009	2329	5.15	0.47	1.1
2010	2573	5.56	0.86	2.2
2011	2531	5.29	0.71	1.8

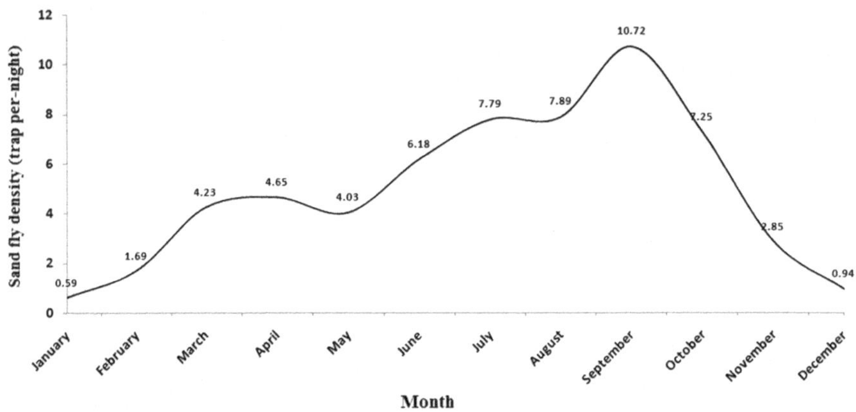

Fig. 6.3 Monthly distribution of average sand fly density during the period from 2005 to 2011 in Muzaffarpur district, Bihar

average maximum relative humidity (in the morning) of this region was recorded as 84.53%, while, in the evening it was documented as 54.21%. The district received maximum average relative humidity (in the morning) in January (89.29%). However, the maximum average relative humidity (in the evening) was recorded in July 72.14%, and the lowest in April (35.94%) during the study period. During the study period, highest rainfall (698.1) was received in 2007. The analysis of the result showed that average annual rainfall in Muzaffarpur district was 102.24 mm (Table 6.2). It reaches its peak in July (291.05 mm), the monsoon month of the year; and the minimum rainfall was recorded in December (0.77 mm). There is generally very less precipitation observed during December to March.

Wind velocity is another important parameter of climatic characteristics, ranging from 0 to 368.5 kmph in Muzaffarpur district during the study period. However, maximum velocity was recorded in July 2006 (368.5 kmph); whereas, minimum velocity was found mostly in the December month.

Table 6.2 Descriptive characteristics of climatic variables of the Muzaffarpur district

Variables	Sand fly density	Temperature (°C)		Rainfall (mm)	Relative humidity (RH) (%)		Wind velocity (km/h)
		Maximum	Minimum		RH_1 (%)	RH_2 (%)	
Average	4.90	30.44	19.65	102.24	84.53	54.21	9.41
Standard deviation	3.44	0.53	0.73	16.78	0.74	1.51	4.38
Kurtosis	−0.64	−0.32	−1.33	4.61	3.22	1.39	80.18
Skewness	0.52	−0.72	−0.43	2.12	−1.70	−0.61	8.87

RH_1 Relative humidity (in the morning); *RH_2* Relative humidity (in the evening)

Association between meteorological variables and Sand fly density

The Pearson correlation coefficient was used to estimate the association between sand fly density and the recorded climatic variables from the IMD station during the period of 2005–2011. Table 6.3 showed an independent association of potential climatic variables with sand fly density in each year. A significant and strong association existed between the sand fly density and maximum temperature ($r = 0.57$; $P < 0.05$), minimum temperature ($r = 0.76$; $P < 0.007$) and moderate association was observed with the relative humidity in evening ($r = 0.45$; $P < 0.002$) and rainfall ($r = 0.48$; $P < 0.05$). However, when the monthly numbers of sand fly density and the recorded temperature, relative humidity, rainfall, wind velocity during the period from 2005 to 2011 were compared each year separately, there was a consistent significant correlation was found between sand fly density and maximum temperature, minimum temperature, relative humidity (evening) and rainfall. The correlation coefficients of these relationships between the variables are shown a strong and positive correlation existed between sand fly density and wind velocity. Results showed a negative and a weak relationship with sand fly density and RH (morning).

Association between meteorological variables and kala-Azar incidences

Table 6.3 represents the Pearson correlation coefficients of VL incidence with respect to climatic variables. When the monthly numbers of reported cases were compared each year separately, there was a significant association was established between maximum temperature and cases incidence ($r = 0.61$; $P < 0.032$), however a moderate to weak association relationship was found with the minimum temperature ($r = 0.36$), except in 2006. In our result, a very poor and insignificant correlation was observed between the VL incidences and rainfall ($r = 0.096$; $P < 0.75$) during the study period. The correlation coefficients of between RH_1 and disease incidence showed a strong and negative correlation ($r = -0.66$; $P < 0.005$); while poor correlations were evidenced with the RH_2, except in 2011. The

Table 6.3 Pearson correlation coefficients of visceral leishmaniasis (VL) incidence vis-à-vis sandfly density with respect to climatic variables

Year	Variables	Temperature (°C)		Relative humidity (%)		Rainfall (cm)	MWS
		Maximum	Minimum	RH_1	RH_2		
2005	Case	0.49#	0.25	−0.33	−0.34*	0.05	0.39
	Sand fly density	0.68*	0.88**	−0.23	0.67*	0.65*	0.58*
2006	Case	0.74*	0.54*	−0.78*	0.01	0.07	0.34
	Sand fly density	0.66*	0.75*	−0.23	0.13	0.45#	0.35
2007	Case	0.58*	0.46#	−0.51#	−0.03	0.06	0.001
	Sand fly density	0.47*	0.67*	0.32	0.63	0.57#	0.36
2008	Case	0.36*	0.04#	−0.23#	−0.70**	−0.22	0.45*
	Sand fly density	0.58*	0.82**	−0.17	0.50#	0.67*	0.74**
2009	Case	0.43*	0.15#	−0.27#	−0.70**	−0.19#	0.40*
	Sand fly density	0.61*	0.75**	0.06	0.58*	0.56#	0.47#
2010	Case	0.69*	0.41	−0.55#	−0.29	0.23	0.45#
	Sand fly density	0.57*	0.78**	0.07	0.55#	0.49#	0.28
2011	Case	0.58#	0.16	−0.86**	−0.59*	0.07	0.44#
	Sand fly density	0.45#	0.72*	0.31	0.56*	0.57*	0.23

Significance level—#$P < 0.10$; *$P < 0.05$; **$P < 0.01$

association between MWS and monthly distribution of Kala-Azar cases showed moderate significant association (r = 0.41; $P < 0.009$), apart from the year 2007.

Cluster-outlier (CO-type) analysis

Local Moran's *I* results generally concur as to the areas with the greatest clustering of VL incidence for all of the years combined (2005–2011). The cluster map was obtained with <0.01 significance level (Fig. 6.4). The Moran's *I* for the period between 2005 and 2011 is calculated as 0.52 ($P < 0.01$) while the z-score is estimated as 24.06 ($P < 0.01$). The analysis determined that the null hypothesis is rejected because the spatial distribution of VL cases since 2005–2011 in the district is more spatially clustered than would be expected if the underlying spatial process were random. Villages that formed high density clusters were evidenced in gray, while villages with low density clusters were evident in blue. High-high clusters covered a wider area in the western part of the district. Some small clusters were also generated in the southern and central part of the district. The low-low clusters were observed in the central and eastern part of the district.

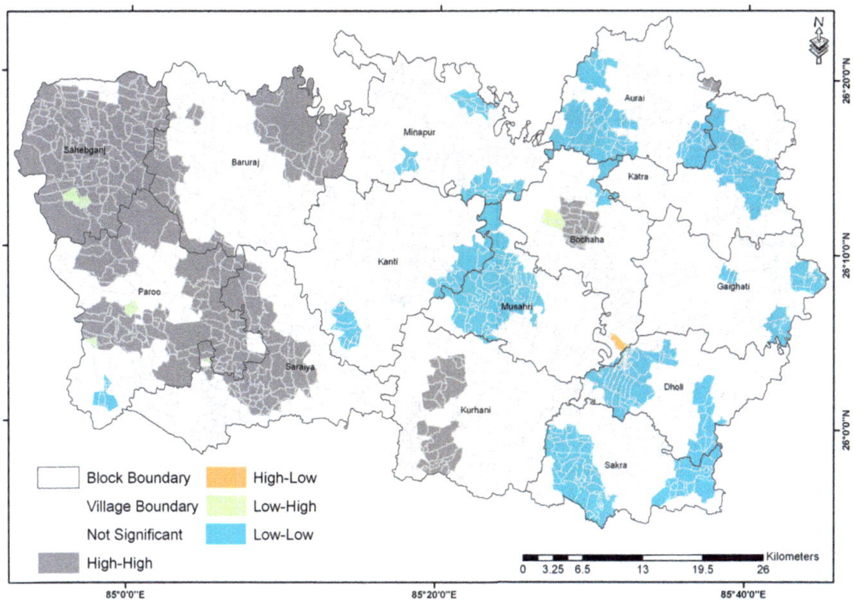

Fig. 6.4 Spatial clustering pattern of VL in Muzaffarpur district during period between 2005 and 2011

Hot spot and cold spot analysis

Results from *Getis-Ord Gi** analysis showed that G_i^* index is 0.31 ($P < 0.01$). From this result, we can expect the VL cases pattern in Muzaffarpur district is exhibiting a cluster pattern. The z-score for VL incidence within the district is 8.64 ($P < 0.01$). With the z-score, there will be small probability which is less than 1% likelihood that the clustering of high values could be a result of random chance; so, we can reject the null hypothesis. Figure 6.5 depicts the hotspot (i.e., clustered of high VL incidence village) and cold spot (i.e., clustered of low VL incidence village) areas in Muzaffarpur district from 2005 to 2011, based on *Getis-Ord Gi** statistic in ARC GIS software v9.3. Two wider hot spot areas were demarcated in the western part, located in in Paroo and Sahebganj block. Another small pocket of hotspot areas were marked in the northern part of the district (Baruraj block), Bochaha and Kurhani block, located in central and southern part of the district respectively. On the other hand, cold spots areas were found in the central part of the Muzaffarpur district (Musahari block) which is located in and around the urban and peri-urban region. Some cold spot were also observed in the northeast (Aurai and Katra) and southeast corner of the study site (Sakra and Dholi block) and eastern part (Gaighata block).

Probable area of high VL incidence

Finally a high VL risk areas was demarcated based on the cumulative case incidence per 10,000 populations from 2005 to 2011 (Fig. 6.6). The overall Root mean

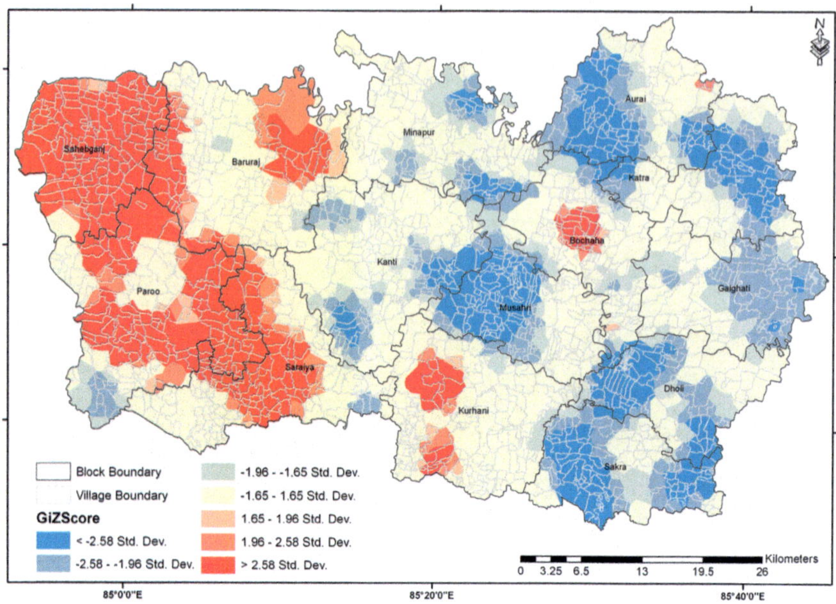

Fig. 6.5 Hot spot and cold spot areas of Visceral Leishmaniasis in Muzaffarpur district during the period from 2005 to 2011

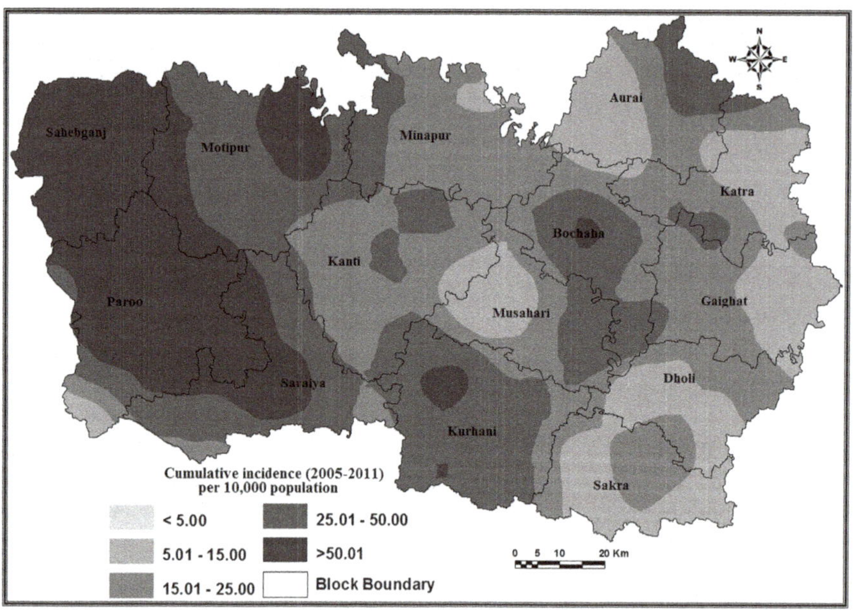

Fig. 6.6 Visceral Leishmaniasis (VL) risk zone map of Muzaffarpur district during the period from 2005 to 2011

square error (RMSE) of the ordinary krigging model is recorded as 0.13. Results showed Sahebganj and Paroo block are at higher risk (VL incidence rate is more than 50/10,000 population). Some higher small pockets were also exemplified in the northern part of Motipur, Saraiya and Aurai blocks. Low VL incidence areas were observed in the central part of Musahari PHC, which is completely urban and peri-urban. In the eastern and southeast region, the incidence rate was somewhat low. However, Motipur, Kurhani and Central part of Bochaha blocks portrayed the medium risk for VL incidence.

6.5.4 Discussion

The VL data analyzed in this paper correspond to cases treated in public health facilities and some Non-government Organizations (NGOs) in Muzaffarpur district during 2005–2011. During the study period, the VL incidence rate increased from 2005 onwards, and was reduced from 2009. However this phenomenon may simply reflect an increase of the proportion of reported cases linked to the launch of the VL elimination initiative and an intensification of control efforts in 2005 (Kumar et al. 2011; Mondal et al. 2009). Additionally, the endorsement of public health facilities for VL treatment was started since 2006 when monetary incentives were provided to patients attending health centers in Bihar. It is noteworthy that the underreporting of VL decreased from 2003 to 2006 in Bihar (Singh et al. 2010) and the number of cases reported by NGOs increased during this period. Furthermore, the spraying techniques and upgrading of household characteristics has been undertaken for effective vector control with the involvement of the community and the health service infrastructure (Kesari et al. 2010). Application of environmental management with modification aiming at reducing and controlling VL proved to be more efficient than the application of pesticides (Bhunia et al. 2012a, b).

The link between VL and weather parameters has long been the subject to study in the Indian sub-continent. Weather parameters played an important role in preventing the spread of VL (Kuhn 1999). This study also illustrated that the maximum temperature has strong and positive influence on the VL transmission. Summer temperatures in this region are high in the warmest months, and the parasites developed in a certain range of temperature (Killick-Kendrick 1983), and can be transmitted if the suitable vector is present. However, a strong association was delineated between the maximum and minimum temperature with the sand fly density, and also ratified to the previous studies (Picado et al. 2010; Ozbel et al. 2011; Bhunia et al. 2012a, b). Napier et al. (1926) and Singh et al. (2008) suggested that RH is the most critical determinants for sand fly survival and disease transmission. Our results showed that negative and poor correlation between sand fly density and RH_{-1}; although, significant correlation was generated with the evening RH_2. It may be due to newly laid eggs are subject to desiccation and the adults to moisture-related reductions in survival throughout their lifetimes. The results of our study also showed not any significant relationship between the mean wind velocities with the monthly incidence of VL, whereas strong relationship was found with the sandfly density. It may be due to the wind tends to suppress sand fly flight; on the other hand, it could have affected their oviposition by offering fertile ground for speculation. This effect could be attributed to reducing sand fly abundance due to higher wind velocity. In this present study, not any significant connection was generated between the monthly Kala-Azar incidences and wind velocity. However, our result showed positive and significant association with the monthly sand fly abundance and rainfall. Rainfall may affect sub-surface soil moisture that intensifies the sand fly breeding. In addition, heavy rainfall or storms may destroy existing breeding sites, interrupt the development of sand fly eggs or larvae, or simply flush the eggs or larvae out of the pools (Mukhopadhyay et al. 1990).

Present study also found out the VL incidence villages in the district were spatially autocorrelated and their pattern of distribution were clustered rather than random chances. This was supported by Local Moran's I statistics. Using *Getis-Ord Gi** statistic the hotspot of cold spot of VL incidence villages were delineated. However, traditional use of incidence data cannot discriminate the spatial differences in threat areas from those of their neighboring areas, which decisive to the effective control of vector-borne infectious diseases (Wen et al. 2010). These information may assist health agencies, epidemiologist, public health officer, and responsible authorities to combat VL incidences. However, GIS provided valuable information as well as the capability to find out spatial liaison among spatial location of VL cases through spatial and statistical analysis. On the other hand, the detection of high VL incidence probability region provides the opportunity to investigate conditions responsible for stimulating outbreaks (Woodruff et al. 2002; Brownstein et al. 2002) and allocates for the development of '*early warning systems*' (Hay et al. 2002; Barbazan et al. 2002; Kaninda et al. 2000).

6.5.5 Conclusion

Our study encourages the implementation of weather parameters and spatial statistics as competent tools for VL surveillance and monitoring systems. Temperature, humidity (evening), rainfall, and mean wind velocity have a mammoth importance on sand fly abundance, may helpful to promulgate the disease in this area. Moreover, a cluster investigation of our study makes out high and low significant VL incidence areas could help to understand, at local scale, in VL control programme. Spatial patterns of cumulative VL risk were observed in the Muzaffarpur district revealing micro-high risk areas for VL infection. Conversely, the smoothed map generated through the interpolation of VL incidence data for last seven years, allowed to make predictions on the VL density in those villages where the monitoring design was not statistically efficient, and, in which the monitoring pattern need to be increased in order to achieve satisfactory results. Spatial analysis using epidemiological apparatuses exposes spatial and temporal clustering of VL cases within the district and their connotation with the meteorological parameters. This evidence can be used in scheming outbursts in the future. This work endorses scope and possibility of geospatial investigation in public health in India.

References

Anselin L (1995) Local indicators of spatial association—LISA. Geograph Anal 27(2):93–115

Assimina Z, Charilaos K, Fotoula B (2008) Leishmaniasis: an overlooked public health concern. Health Sci J 2(4):196–205

Barbazan P, Yoksan S, Gonzalez J (2002) Dengue hemorrhagic fever epidemiology in Thailand: description and forecasting of epidemics. Microbes Infect 4:699–705

Bayer AM, Hunter GC, Gilman RH, Cornejo del Carpio JG, Naquira C, Bern C, Levy MZ (2009) Chagas disease, migration and community settlement patterns in Arequipa, Peru. PLoS Negl Trop Dis 3(12):e567

Bhunia GS, Chatterjee N, Kumar V, Mandal R, Das P, Kesari S (2012a) Remote sensing and GIS: tools for the prediction of epidemic for the intervention measure. In: 14th annual international conference and exhibition on geospatial international technology and application, 7–9 Feb 2012, Gurgaon, India

Bhunia GS, Chatterjee N, Kumar V, Siddiqui NA, Mandal R, Das P (2012b) Delimitation of Kala-Azar risk areas in the district of Vaishali in Bihar (India) using a geo-environmental approach Memórias do Instituto Oswaldo Cruz, vol 107(5), pp 609–620

Bhunia GS, Kesari S, Chatterjee N, Kumar K, Das P (2013) Spatial and temporal variation and hotspot detection of Kala-Azar disease in Vaishali district (Bihar), India. BMC Infect Dis 13:64. https://doi.org/10.1186/1471-2334-13-64

Bhunia GS, Kesari S, Jeyaram A, Kumar V, Das P (2010a) Influence of topography on the endemicity of Kala-Azar: a study based on remote sensing and geographical information system. Geospatial Health 4(2):155–165

Bhunia GS, Kumar V, Kumar AJ, Das P, Kesari S (2010b) The use of remote sensing in the identification of the eco–environmental factors associated with the risk of human visceral leishmaniasis (Kala-Azar) on the Gangetic plain, in north–eastern India. Ann Trop Med Parasitol 104(1):35–53

Bobadilla Suarez M, Ewen JG, Groombridge JJ, Beckmann K, Shotton J, Masters N, Sainsbury AW (2017) Using qualitative disease risk analysis for Herpetofauna conservation translocations Trans Ecol Geograph Barriers Ecohealth 14(Suppl 1), 47–60. http://doi.org/10.1007/s10393-015-1086-4

Bouchard C, Leighton PA, Beauchamp G, Nguon S, Trudel L, Milord F, Lindsay LR, Bélanger D, Ogden NH (2013) Harvested white-tailed deer as sentinel hosts for early establishing Ixodesscapularis populations and risk from vector-borne zoonoses in southeastern Canada. J Med Entomol 50(2):384–393

Brownstein JS, Rosen H, Purdy D, Miller JR, Merlino M, Mostashari F, Fish D (2002) Spatial analysis of West Nile virus: rapid risk assessment of an introduced vector-borne zoonosis. Vector Borne Zoonotic Dis 2:157–164

Cliff AD, Hord JK (1981) Spatial processes: models and applications. Pion, London, UK

Croner CM, De Cola L (2001) Visualization of disease surveillance data with geostatistics. In: Presented at UNECE (United Nations Economic Commission for Europe) work session on methodological issues involving integration of statistics and geography, Tallinn. http://www.unece.org/stats/documents/2001/09/gis/25.e.pdf

Dhiman RC, Yadav RS (2016) Insecticide resistance in phlebotomine sandflies in Southeast Asia with emphasis on the Indian subcontinent. Infect Dis Poverty 5:106. https://doi.org/10.1186/s40249-016-0200-3

Eisen L, Eisen RJ (2011) Using geographic information systems and decision support systems for the prediction, prevention, and control of vector-borne diseases. Ann Rev Entomol 56:41–61

Fotheringham AS, Brunsdon C, Charlton ME (2002) Geographically weighted regression: the analysis of spatially varying relationships. Wiley, Chichester

Frössling J, Ohlson A, Björkman C, Håkansson N, Nöremark M (2012) Application of network analysis parameters in risk-based surveillance: examples based on cattle trade data and bovine infections in Sweden. Prev Vet Med 105(3):202–208

Gething PW, Noor AM, Gikandi PW, Ogara EAA, Hay SI, Nixon MS, Snow RW, Atkinson PM (2006) Improving imperfect data from health management information systems in africa using space-time geostatistics. PLoS Med 3:825–831

Getis A, Ord JK (1992) The analysis of spatial association by use of distance statistics. Geogr Analy 24(3):189–206

Guerin PJ, Olliaro P, Sundar S, Boelaert M, Croft SL, Desjeux P, Wasunna MK, Bryceson ADM (2002) Visceral leishmaniasis: current status of control, diagnosis, and treatment, and a proposed research and development agenda. Lancet Infect Dis 2:494–501

Hay SI, Simba M, Busolo M, Noor AM, Guyatt HL, Ochola SA, Snow RW (2002) Defining and detecting malaria epidemics in the highlands of Western Kenya. Emerg Infect Dis 8:555–562

Hillesland H, Read A, Subhadra B, Hurwitz I, McKelvey R, Ghosh K, Das P, Durvasula R (2008) Identification of aerobic gut bacteria from the Kala Azar Vector, Phlebotomus argentipes: a platform for potential paratransgenic manipulation of sand flies. Am J Trop Med Hyg 79 (6):881–886

Hu J, Liu Y, Wimberly MC (2014) FDEOD—A software framework for downloading earth observation data. In: Proceedings of the Association of Computing Machinery Southeast Conference, Kennesaw, GA

Istudor N, Ursacescu M, Sendroiu C, Radu I (2016) Theoretical framework of organizational intelligence: a managerial approach to promote renewable energy in rural economies. Energies 9(639):1–20. https://doi.org/10.3390/en9080639

Kaninda A, Belanger F, Lewis R, Batchassi E, Aplogan A, Yakoua Y, Paquet C (2000) Effectiveness of incidence thresholds for detection and control of meningococcal meningitis epidemics in northern Togo. Int J Epidemiol 29:933–940

Kesari S, Bhunia GS, Chatterjee N, Kumar V, Mandal R, Das P (2013) Appraisal of Phlebotomus argentipes habitat suitability using a remotely sensed index in the Kala-Azar endemic focus of Bihar. India Memórias Do Instituto Oswaldo Cruz 108(2):197–204. https://doi.org/10.1590/0074-0276108022013012

Kesari S, Bhunia GS, Kumar V, Jeyaram A, Ranjan A, Das P (2010) Study of house-level risk factors associated in the transmission of Indian Kala-Azar. Parasit Vectors 3:94. https://doi.org/10.1186/1756-3305-3-94

Killick-Kendrick (1983) Investigation of Phlebotomine sandflies. In: Biology of the kinetoplastide, vol 2, Lumsden WHR, Evans DA (eds) London, Academic Press

Kilpatrick AM, Pape WJ (2013) – Predicting human West Nile virus infections with mosquito surveillance data. Am J Epidemiol 178(5):829–835

Kleinschmidt I, Bagayoko M, Clarke GPY, Craig M, Sueur DL (2000) A spatial statistical approach to malaria mapping. Int J Epidemiol 29:355–61

Kuhn KG (1999) Global warming and leishmaniasis in Italy. Bull. Trop. Med. Int. Hlth. 7:1–2

Kumar N, Singh SP, Mondal D, Joshi A, Das P, Sundar S, Kroeger A, Hirve S, Siddiqui NA, Boelaert M (2011) How do health care providers deal with Kala-Azar in the Indian subcontinent? Indian J Med Res 134:349–355

Kumar V, Shankar L, Kesari S, Bhunia GS, Dinesh DS, Mandal R, Das P (2015) Insecticide susceptibility of Phlebotomus argentipes & assessment of vector control in two districts of West Bengal, India. The Indian Journal of Medical Research 142(2):211–215. https://doi.org/10.4103/0971-5916.164260

Lainson R (1988) Ecological interactions in the transmission of the leishmaniases. Philos Trans Royal Soc London Ser B, Biol Sci 321(1207):389–404

Lang L (1998) GIS-a proven tool for public health analysis. J Environ Health 67

Medenica S, Jovanović S, Dožić I, Milicić B, Lakićević N, Rakocević B (2015) Epidemiological Surveillance of Leishmaniasis in Montenegro, 1992–2013. Srpski arhiv za celokupno lekarstvo 143(11–12):707–711

Medford-Davis LN, Kapur GB (2014) Preparing for effective communications during disasters: lessons from a World Health Organization quality improvement project. Int J Emerg Med 7:15. https://doi.org/10.1186/1865-1380-7-15

Métras R, Collins LM, White RG, Alonso S, Chevalier V, Thuranira-McKeever C, Pfeiffer DU (2011) Rift Valley fever epidemiology, surveillance, and control: what have models contributed? Vector Borne Zoonotic Dis 11(6):761–771

Mitchell A (2005) The ESRI guide to GIS analysis, vol 2. ESRI Press

Mondal D, Singh SP, Kumar N, Joshi A, Sundar S, Das P, Hirve S, Kroeger A, Boelaert M (2009) Visceral Leishmaniasis elimination programme in India, Bangladesh, and Nepal: reshaping the case finding/case management strategy. PLoS Negl Trop Dis 3(1):e355. https://doi.org/10.1371/journal.pntd.0000355

Mukhopadhyay AK, Rahman SJ, Chakravarty AK (1990) Effects of flood control on immature stages of sandflies in flood prone Kala-Azar endemic villages of North Bihar, India. WHO/VBC 90:986

Musa GJ, Chiang PH, Sylk T, BavleyR, Keating W, Lakew B, Tsou HC, Hoven CW (2013) Use of GIS mapping as a public health tool—from cholera to cancer. Health Serv Insights 6:111–116

Napier LE (1926) An epidemiological consideration of the transmission of Kala-Azar in India. India Med Res Memoir 4:219–265

Njabo KY, Cornel AJ, Sehgal RNM, Loiseau C, Buermann W, Harrigan RJ, Pollinger J, Valkiūnas G, Smith TB (2009) Coquillettidia (Culicidae, Diptera) mosquitoes are natural vectors of avian malaria in Africa. Malaria Journal 8:193. https://doi.org/10.1186/1475-2875-8-193

Ozair M, Lashari AA, Jung IH, Okosun KO (2012) Stability analysis and optimal control of a vector-borne disease with nonlinear incidence. Discr Dynam Nat Soc 2012, Article ID 595487, p 21 https://doi.org/10.1155/2012/595487

Ozbel Y, Sanjoba C, Alten B, Asada M, Depaquit J, Matsumoto Y, Demir S, Siyambalagoda RRMLR, Rajapakse RPVJ, Matsumoto Y (2011) Distribution and ecological aspects of sand fly (Diptera: Psychodidae) species in Sri Lanka. J Vect Ecol 36(1):S77–S86

Picado A, Das ML, Kumar V, Dinesh DS, Rijal S, Singh SP, Das P, Coosemans M, Boelaert M, Davies C (2010) Phlebotomus argentipes seasonal patterns in India and Nepal. J Med Entomol 47(2):283–286

Rajabi M (2015) Disease susceptibility mapping using spatial modeling techniques. Lund University. http://www.nateko.lu.se/sites/nateko.lu.se/files/mohammadreza-rajabi-first_year_-seminar-17-08-2015-v4.pdf

Reist M, Jemmi T, Stärk KDC (2012) Policy-driven development of cost-effective, risk-based surveillance strategies. Prev Vet Med 105(3):176–184

Riner ME, Cunningham C, Johnson A (2004) Public health education and practice using geographic information system technology. Public Health Nurs 21(1):57–65

Ryan KJ, Ray CG (eds) (2004) Sherris medical microbiology (4th ed). McGraw Hill ISBN 0-8385-8529-9, pp 749–754

Santos-Reyes J, Beard AN (2013). Information communication technology and a systemic disaster management system model. https://pdfs.semanticscholar.org/0c00/0405-1c332b26b046a6a779377b5b21dd006e.pdf

Shah NH, Gupta J (2013) SEIR model and simulation for vector borne diseases. Appl Math 4:13–17

Singh A, Roy SP, Kumar R, Nath A (2008) Temperature and humidity play a crucial role in the development of P. *argentipes*. J Ecophysiol Occupation Health 8(1 & 2)

Singh VP, Ranjan A, Topno RK, Verma RB, Siddique NA, Ravidas VN, Kumar N, Pandey K, Das P (2010) Estimation of under-reporting of visceral leishmaniasis cases in Bihar. India Am J Trop Med Hyg 82(1):9–11

Srividya A, Michael E, Palaniyandi M, Pani SP, Das PK (2002) A geostatistical analysis of the geographic distribution of lymphatic filariasis prevalence in southern India. Am J Trop Med Hyg 67:480–489

Sudhakar S, Srinivas T, Palit A, Kar SK, Bhattacharya SK (2006) Mapping of risk prone areas of Kala-Azar (Visceral leishmaniasis) in parts of Bihar state, India: an RS and GIS approach. J Vect Borne Dis 43:115–122

Sutherst RW (2004) Global change and human vulnerability to vector-borne diseases. Clin Microbiol Rev 17(1):136–173. https://doi.org/10.1128/CMR.17.1.136-173.2004

Sutherst RW (2000) Climate change and invasive species—a conceptual framework. In: Mooney HA, Hobbs RJ (eds) Invasive species in a changing world. Island Press, Washington, D.C, pp 211–240

Swaddle JP, Calos SE (2008) Increased avian diversity is associated with lower incidence of human West Nile infection: observation of the dilution effect. PLoS One 3(6):e2488

Thakur CP (2007) A new strategy for elimination of Kala-Azar from rural Bihar. Indian J Med Res 126:447–451

Thompson PN, Etter E (2015) Epidemiological surveillance methods for vector-borne diseases. Rev Sci Tech Off Int Epiz 34(1):235–247

Van der Kelen P, Downs JA, Unnasch T, Stark L (2014) A risk index model for predicting eastern equine encephalitis virus transmission to horses in Florida. Appl Geogr 48:79–86

Wang H, Liang G (2015) Epidemiology of Japanese encephalitis: past, present, and future prospects. Ther Clin Risk Manag 11:435–448. https://doi.org/10.2147/TCRM.S51168

Wang J, Cao Z, Zeng DD, Wang Q, Wang X, Qian H (2014) Epidemiological analysis, detection, and comparison of space-time patterns of Beijing hand-foot-mouth disease (2008–2012). PLoS One 9(3):e92745. https://doi.org/10.1371/journal.pone.0092745

Wen TH, Lin NH, Chao DY, Hwang KP, Kan CC, Lin K et al (2010) Spatial-temporal patterns of dengue in areas at risk of dengue hemorrhagic fever in Kaohsiung, Taiwan, 2002. Int J Infect Dis 14(4):e334–e343

Woodruff RE, Guest CS, Garner MG, Becker N, Lindesay J, Carvan T, Ebi K (2002) Predicting Ross river virus epidemics from regional weather data. Epidemiology 13:384–393

World Health Organization (WHO) (2015) Kala-Azar elimination programme: report of a WHO consultation of partners, Geneva, Switzerland, 10–11, February 2015. http://apps.who.int/iris/bitstream/10665/185042/1/9789241509497_eng.pdf

World Health Organization (WHO) (1998) Life in twenty first century: a vision for all. World Health Report, World Health Organization, Geneva, Switzerland

Yoo EH (2014) Site-specific prediction of West Nile virus mosquito abundance in greater Toronto Area using generalized linear mixed models. Int J Geogr Inf Sci 28(2):296–313

Chapter 7
Spatial Technology in Health-Care Planning and Policy

7.1 Introduction

Geography and health are inherently connected. Although health geography is diligently united with epidemiology its distinctive primary accent is on the spatial associations and configurations. Whereas epidemiology is forecast on the biomedical model and emphasis on the ecology of disease, health geography pursues to reconnoiter the social, cultural and political circumstances for health within an agenda of spatial association. There is a necessity to apprehend disease menace aspects and how threats such as life style, genetics, occupation and environment interrelate with the social and natural environments (Dummer 2008). Understanding geography comprising the procedure of health services and the situation and nature of environmental disclosures is decisive in evaluating the interrelations innate in several health related risk exposure (Dummer 2008). Concerned with the scale of organization, worldwide complications such as environmental change, demographic transition and internationalization of health service organization all have geographic framework that openly effect health policy.

Researches to date have employed variables from current satellite technologies that were not intended to support investigation on infectious disease ecology. Such data have been improved to assist as representations for pertinent signs that might characterize significant risk indicators. On the other hand geographic information system (GIS) sustenance public health in various ways containing mapping, monitoring and modelling infectious and chronic diseases, diseases reconnaissance and epidemic recognition, emergency preparedness, and pursuing interventions and health promotions creativities. Depending on the emerging infectious disease of concern, climate, land use pattern, agricultural practices, deforestation, extent of water bodies as well as human habitation may all influence on risk of particular infectious disease. The spatial logical methods and hardware/software forecasts to reconnoiter such enormous data sets are just now being established. Certain approaches will carry over with tiny amendment from digital image processing.

G. S. Bhunia and P. K. Shit, *Geospatial Analysis of Public Health*,
https://doi.org/10.1007/978-3-030-01680-7_7

However, it is mainly concern to attain pertinent temporal and spatial scale is precarious. Once such epidemiological data are assembled, the contest must be lectured of coalescing them with appropriate tools and acquaintance from other chastisements such as spatial analysis, medical geography, infectious disease epidemiology, sociology, international relations and economics (Wilson 2002).

Growing health informatics systems promise to transfigure health and healthcare programmes worldwide. Hesse et al. (2010) provide a valuable outline for examining health informatics technologies, under the common term "Health 2.0" in three domains as, personal health information system, clinical heath information system and public health information system. Personal health information services enable patients by permitting them to observe their food, exercise programmes, lifestyle changes, non-prescription treatments, and nutritional appendages. The clinical health information system concerned about plentiful medications and professional recommendations evolving from research. Consequently, the public health information system disquieted about the information composed by the US national center for Health Statistics, Centers for Disease Control, Census, Health and Human Services, the World Health Organization and other agencies provide valuable information to guide public policy. The data are integrally geo-spatial and temporal enriching contests of funding patterns, associations, clusters, gaps and outliers.

7.2 Environment and Space Technology in Public Health Planning and Policy

Recent advances the spatial and spectral resolutions of satellite image data and frequency of satellite image, united with new methods and software are growing signifying prospects to smear such data to difficulties in which contact, pattern distances, or environmental factors are imperative. Most of the research work recommended that certain environmental characteristics significant to acquaintance or transmission are not completely seized by coarse spatial resolution data, and that finer resolution, as well as better spectral resolution and details, should improve research and deliver more valuable consequences. The high spatial resolution, microwave and hyperspectral satellite data increase our appreciative of the epidemiology, transmission patterns and risk factors for such diseases.

Low spatial resolution images have been used in studies of various kinds of diseases where environmental exposures are significant. Satellite data and spatial analysis have been revealed beneficial in non-infectious disease valuation concerning pesticide exposure (Ward et al. 2000). Several reports have been published that are used lower spatial resolution images in studies of infectious disease. However, high spatial resolution data and analytical tools might enhance such research by delineating the relevant environmental characteristics of transmission, diverse characteristics of infectious disease, greater characterization of environmental properties using enhanced spatial and spectral resolution data and by improving the tools for managing and analyzing these data available and easily applied.

Most vector-borne diseases are prejudiced by ecological characteristics that can be perceived openly or circuitously from satellite images. Remotely sensed information deliberated valuable in accepting these diseases comprised configurations of vegetation (crop type, deforestation) habitat type (forest patches, ecotones), fresh water sources (permanent water, wetlands, flooding, soil moisture, canals), housing (human settlements, urban features), and ocean circumstances (ocean color, sea surface temperatures, sea surface height). For instance, satellite derived vegetation appearances (e.g., amount of forest, ecotone or crops) were employed to appraise diseases as various as malaria, leishmaniasis, chagas disease, hantavirus, lyme disease, plague, rift valley fever, schistosomiasis, trypaosomiasis and yellow fever. None of these microbes affecting these diseases is governed by vegetation for their survival or transmission (Table 7.1). Correspondingly, water-related characteristics (e.g., permanent tor seasonal water body, extent of wetlands, flooding, and presence of canals) acquired from satellite images were also employed to investigate numerous infectious diseases as well as cholera, filariasis and onchocerciasis. Thus there is no simple algorithm to outline which remotely-sensed ecological physiognomies will deliver the most insight for any specific disease.

Earlier works primarily concerned to detect connotations between spatial patterns of environmental structures and vector profusion or disease incidence. These relations have been employed to generate 'risk maps' of improved transmission prospect (Kitron 2000). This is predominantly correct for risk maps consequent from epidemiological data that area spatially referenced to enormous areas, but also

Table 7.1 Examples of environmental effects on transmission of different types of microbes classified according to their mode of transmission and reservoir (*Source* Wilson 2002)

Disease and transmission types	Cycle components	Examples of direct environmental determinants	Examples of indirect environmentally-based pathways
Anthroponoses—direct	Human → Human	Few (Temperature?, Humidity?)	Body fluids or excrement aerosols, human behavior
Anthroponoses—indirect	Human → Vector → Human	Soil, water, vegetation, humidity, wind, food and temperature	Vector abundance and behaviors, cropping patterns, sanitation, irrigation, urbanization, human behavior
Zoonoses—direct	Animal → Animal → Human	Vegetation, land cover, humidity, temperature	Animal reservoir abundance, human recreation and behavior, domestic animal husbandry, pet owners patterns
Zoonoses—indirect	Animal → Vector → Human	Soil, water, vegetation, humidity, wind, food, temperature	Animal reservoir abundance, vector abundance, crop types urbanization, wildlife habitat, housing type, human recreation and behavior, domestic animal husbandry, pet ownership pattern

for those that extrapolate from vector or reservoir abundance data. Thus the expediency of remotely sensed data in recitation environmental relations with disease risk somewhat depends on the spatial scale of the obtainable disease-indicator data. Several of infectious diseases, yet, have local, habitat-specific causes of risk. Research intended at accepting mechanisms, and ultimately dropping that risk should be able to assistance from extensive coverage of transmission-specific ecological variables that remotely sensed data provides. Once such relations are rationally well recognized, remotely sensed data offer an even superior utensil for undertaking environmental reconnaissance that might lead to early warning of epidemics or location of elevated risk.

Current advances in health geography comprise a stronger importance on health variations and impact of spatial and social downgrading on health, more unambiguous examination of the nature of health change. GIS cares the analysis of vibrant spatial data are being more diligently affiliated with global positioning systems to observe the movement of people in real time to contextualize the interrelations between the physical and built environments (Boothby and Dummer 2003).

7.3 Geographic Research and Public Health-Care Planning and Policy

There is plentiful information that the use of geographic information has had a histrionic influence on the effort of health specialists. Medical epidemiologists, the front line of disease investigators have employed broadly in their bout against infections that have a clear association among person, place and time. Geomedicine spread over the supremacy of GIS technology to individual health. As this concept endures to be espoused by more health care specialists, physicians will progressively use geomedicine to aid diagnose, treat, and avert disease and in various cases even create commendations to patients on where they might sentient, work and play. Tele-geoprocessing is a novel stream circling around real-time spatial database that are simplified frequently by means of telecommunications system so as to maintenance problem solving and decision making at a time and any place (Xue et al. 2002).

The appreciation of increasing importance of public health heartened medical geographers to pursue an indulgent of diseases that had substantial influences on populations in both developing and developed countries that articulated themselves within the interchange of the physical and social atmosphere in which they subsist. There was also the contest of estrangement medical geography from the deception of environmental determinism. With progressively meticulous epidemiological data and quickly enlightening computing power, medical geographers have paid to the spatial modeling at local level to the regional level to space-time modeling of pandemics at the global level. Spatial information management is a discipline for the individual organization, administration, the micro level and for the society in general the macro level. The concept covers various disciplines such as capture, storing maintenance and upgrading of data and information, information technology, organizational issues and spatial data infrastructures. A number of investigations have employed GIS to study disease patterns, spatio-temporal disparities in health outcomes, and recognized probable causes of mapped configurations (Davis et al. 2014). This tool can also be employed to mark resources for disease inhibition by emphasizing areas with considerably high rates and to envisage which zones might be at imminent risk and which may be assistance most from future local population screening.

7.4 Telehelth System and Health-Care Planning and Policy

Rapid developments in information and communications technologies have enabled the wide spectrum of communications technologies suitable for all across the digital divide. Telehealth is by far the most suitable information for the alert and preparedness phase as it makes it possible to predict the outbreak threat based on RS and GIS platforms, and to spread the information quickly to 'those who need to know', so decisions can be made and action taken at the early response level (Souarès 2000). One of the major telehealth applications in disease epidemiology would have contributed to 'monitoring of well-being' based on two important areas viz. disease management (including use of vital signs monitor—notably for uses with long terms conditions) through peer information exchange and emotional support and health and fitness (including use of telephone and video consultation and dialogue about lifestyles, health and motivational coaching of public health workers) (Souarés 2000).

For inhabitants of rural areas, especially in developing countries access to medical specialists are problematic issue (Gupta et al. 2003); since it needs travel to urban centers where the doctors are present. Such travel is often costly and sometimes tricky for rural people. For these places telehealth initiative may be a significant opportunity for specialized care and diagnosis of infectious diease for rural populations and for the evaluation and monitoring of healthcare services. However, telehealth network may provide a good opportunity for the patients to have access to remote hospitals for real time evaluation by specialists, via either video or audio, using Doppler's and webcams (Bhunia et al. 2012).

Vector control can be a useful approach to reduce the incidence of any vector-borne disease. The diversity of vector biology is such that there is no one method to control either applications or attainable in the different foci throughout the world. The principle methods used in vector control have been the application of insecticides, sometimes with environmental management, but there has been little emphasis on methods for protection of individuals. The breeding sites of the vector are not known which therefore limits control to attacking of adults. However, they are not necessarily presented in order of importance. Telehealth may help to co-ordinate the strategies to bridge the gap between the people and programmer by tele-education and tele-training. Video conferencing is the most common form of transferring the information to a caregiver.

Telehealth through satellites can play a role in the diagnosis and treatment of urgent patients in the public health management (Stroetmann et al. 2010; Schlachta-Fairchild et al. 2008). However, telehealth may simplify the health decision-making process or communication between healthcare providers and individuals on prevention, diagnosis or management of a health condition. To seek the opinion of a specialist using tele-communication, doctors can connect to the

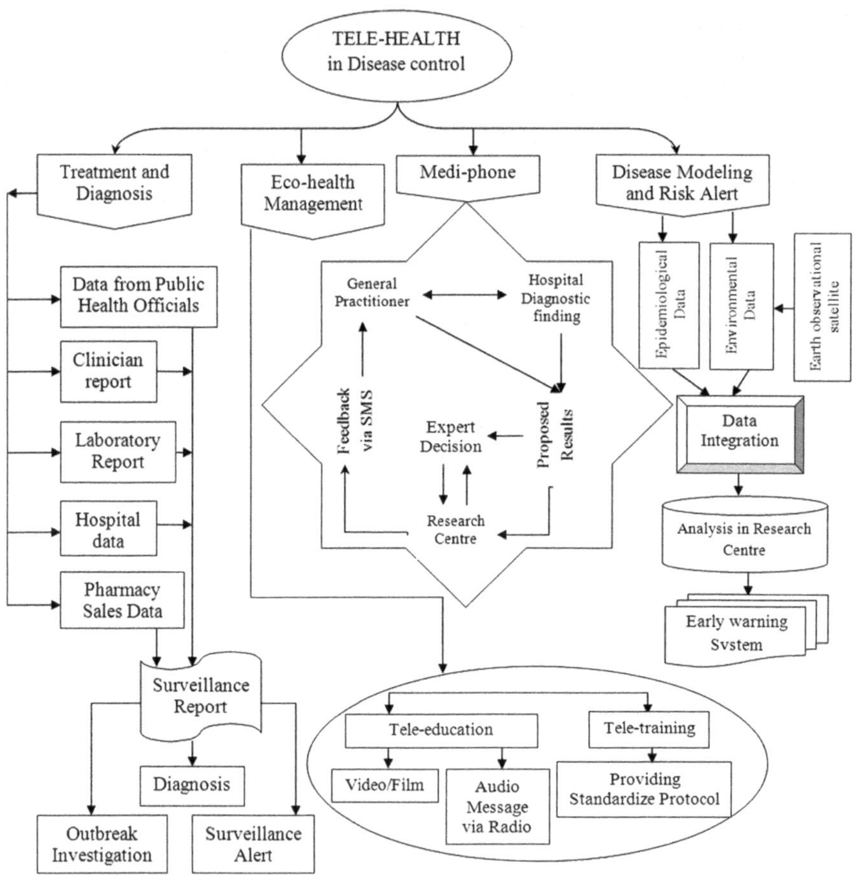

Fig. 7.1 Application of telehealth in disease surveillance system. *Source* Modified after Bhunia et al. (2012)

specialist's personnel computer (PC) from within the telehealth software. The rapid access to a wider range of specialists and medical procedures, to manage overall medical systems and patients care through seamless delivery of service by satellites can provide chances for both patients and medical staff to various urgent medical care treatments. The subject is whether the telehealth method can synergistically assimilate multidimensional intercession, prevention, assessment and management to realize better personnel as well as local, national and international community health (Fig. 7.1).

Telehealth seems to be a promising patient management approach which may reduce the financial burden of health care costs. Telehealth has the potential to dramatically reduce unnecessary costs improving quality of care associated with transport and protection, improved access of medical specialists, improved patient outcomes.

Telehealth may now plan new insights into the geographical distribution and gradients of emergent disease prevalence, as well as on population health assessment, by providing valuable information about different populations at risk, based on risk factor profiles. This information complements the decision support system data to introduce spatial-temporal fluctuations of temperature, humidity, precipitation, land use/land cover, vegetation characteristics etc. Remote sensing techniques have been recently used in this field. Hence, the different data sets can be used for geo-statistical modeling, mapping and geographical and epidemiological analysis. Seamlessly, telehealth system will produce longitudinal data and analysis on advancing health information, social reforms, social policies and practice which will contribute to government, NGO, private and community health system.

7.5 Conclusion

Geospatial technology provides a very rich toolbox of devices and technologies that verve far beyond the measly invention of simple maps. This technology also performances as a dominant sign based rehearsal tools for primary problems recognition and deciphering and enable decision making at all levels. On the other hand, although multiple novel spatial statistical and GIS approaches are possibly obtainable, we still essential to unequivocally regulate which methods explicitly should be employed by practitioners for each specific health condition of interest, and whether the projected approaches are cost-effective and ascendable. However, despite all these abilities for geospatial technology, we endure under-utilized in non-strategic tasks, and in a mostly patchy and clumsy way.

References

Bhunia GS, Kesari S, Chatterjee N, Kumar V, Das P (2012) Telehealth: a perspective approach for visceral leishmaniasis (kala-azar) control in India. Pathogens Global Health 106(3):150–158

Boothby J, Dummer TGB (2003) Facilitating mobility? The role of GIS. Geography 88(4): 300–311

Davis GS, Sevdalis N, Drumright LN (2014) Spatial and temporal analyses to investigate infectious disease transmission within healthcare settings. J Hosp Infect 86:227–243

Dummer TJB (2008) Health geography: supporting public health policy and planning. CMAJ 178 (9):1177–1180

Gupta N, Zurn P, Diallo K, Dal Poz MR (2003) Uses of population census data for monitoring geographical imbalance in the health workforce: snapshots from three developing countries. Int J Equity Health 2:11

Hesse BW, Hansen D, Finholt T, Munson S, Kellogg W, Thomas JC (2010) Social participation in health 2.0. IEEE Comput 43(11):45–52

Kitron U (2000) Risk maps: transmission and burden of vector-borne diseases. Parasitol Today 16 (8):324–325

Schlachta-Fairchild L, Elfrink V, Deickman A (2008) Patient safety, telenursing, and telehealth, patient safety and quality: an evidence-based handbook for nurses, vol 3, pp 3–40. http://www.ncbi.nlm.nih.gov/books/NBK2687/pdf/ch48.pdf

Souarès Y (2000) Telehealth and outbreak prevention and control: the foundations and advances of the Pacific Public Health Surveillance Network. Pacific Health Dialog 7(2):11–28

Stroetmann KA, Robinson LKS, Stroetmann KC, McDaid D (2010) How can telehealth help in the provision of integrated care? World health organization. Denmark, Copenhagen

Ward MH, Nuckols JR, Weigel SJ, Maxwell SK, Cantor KP, Miller RS (2000) Identifying populations potentially exposed to agricultural pesticides using remote sensing and a geographic information system. Environ Health Perspect 108(1):5–12

Wilson ML (2002) Emerging and vector-borne diseases: role of high spatial resolution and hyperspectral images in analyses and forecasts. J Geograph Syst 4:31–42

Xue Y, Cracknell AP, Guo HD (2002) Telegeoprocessing: the integration of remote sensing, geographic information system (GIS), global positioning system (GPS) and telecommunication. Int J Remote Sens 23:1851–1893

Index

G. S. Bhunia and P. K. Shit, *Geospatial Analysis of Public Health*,
https://doi.org/10.1007/978-3-030-01680-7

Printed by Printforce, the Netherlands